NEW SURVEY OF CLARE ISLAND

Volume 9: Birds

Edited by
Thomas C. Kelly

Acadamh Ríoga na hÉireann
Royal Irish Academy

First published in 2020
by the Royal Irish Academy,
19 Dawson St, Dublin 2.
ria.ie

ISBN 978-1-911479-41-3 (PB)
ISBN 978-1-911479-43-7 (PDF)
ISBN 978-1-911479-44-4 (epub)
ISBN 978-1-911479-45-1 (mobi)

British Library Cataloguing-in-Publication Data.

A catalogue record for this book is available from the British Library.

Typesetting and colour reproduction by Compuscript Ltd, Shannon.

Cover photographs by Richard T. Mills.

Printed in Northern Ireland by W&G Baird Ltd, Antrim.

A NOTE FROM THE PUBLISHER

We want to try to offset the environmental impacts of carbon produced during the production of our books. For the production of this book we will plant 8 trees with Easy Treesie.

The Easy Treesie – Crann Project organises children to plant trees. The aim is to plant a tree for every child in Ireland. A million trees by 2023. This is inspired by and part of the Trillion Tree Campaign, which is a project of Plant-for-the-Planet to plant a trillion trees. This initiative is a development and continuation of the activities of the earlier Billion Tree Campaign, which was instigated by Nobel Laureate Wangari Maathai, who founded the Green Belt Movement in Africa in 1977.

Crann – 'Trees for Ireland' is a membership-based, non-profit, registered charity (CHY13698) uniting people with a love of trees. It was formed in 1986 by Jan Alexander, with the aim of "Releafing Ireland". Its mission is to enhance the environment of Ireland through planting, promoting, protecting and increasing awareness about trees and woodlands.

www.easytreesie.com

easy treesie

Crann
TREES FOR IRELAND

CONTENTS

NOTES ON CONTRIBUTORS

Dick Coombes is national coordinator of the Countryside Bird Survey (CBS). He oversees some 380 sample sites across the country which are surveyed by over 220 participants annually, most of whom are volunteers. He participates in several other surveys including I-WeBS and the Swift Survey and is a regular contributor to *Wings* magazine. Dick has had a lifelong interest in birds and is always especially active in spring and autumn when migration is in full swing.

Robert (Bob) Cussen is the lead author on a number of papers in *New Survey of Clare Island. Volume 3: Marine intertidal ecology*. Following a brief period working for the National Parks and Wildlife Survey in North Mayo, he moved to the north east of England to work in nature conservation with Natural England (and its predecessor English Nature). During his time with Natural England he worked on a number of internationally important sites designated for their bird interest and he now works mainly on restoring and maintaining designated peatland and river habitats.

Rachel E. Hutchinson formerly worked as a cataloguer in the Boole Library, University College Cork. She has a lifelong interest in birds and birdwatching, both at home and abroad.

Usna Keating completed a degree in Zoology in University College Cork in 2009. Following this, he completed a MSc degree in Marine Biology and a MSc degree by research, in a project exploring the habitat associations of terrestrial bird species throughout grassland and bog habitats of Ireland. Usna has worked closely with University College Cork, conservation NGOs and public agencies and now works in the fields of environmental consultancy and environmental education.

Thomas C. Kelly is a semi-retired lecturer in Zoology at University College Cork. He holds a PhD on ticks and viruses at seabird colonies in Ireland; his published research includes collaborative peer-reviewed papers on bird–aircraft interactions, Long Distance Dispersal (LDD), and mathematical models of memory in predator-prey systems, novel virus epidemics in seabird colonies, population dynamics of the woodpigeon and lesser black-backed gull.

Eoin Mc Greal works as a conservation ranger for the National Parks and Wildlife Service covering the South-West Mayo area. He studied aquatic ecology at Galway RTC and Wolverhampton Polytechnic. He has a particular interest in the ecology of Mayo's offshore islands and larger inland lakes.

Patrick Smiddy has had a lifelong interest in Ireland's natural history, especially birds and mammals. He has published widely in national and international journals. He worked for the National Parks and Wildlife Service (1980–2008), was secretary of the Irish Rare Birds Committee (1985–96) and was editor of BirdWatch Ireland journal *Irish Birds* (2010–17). He is on the University College Cork Ornithology Group and was awarded an Honorary MSc in 2000.

Stuart Warner has worked with geographical information systems for over 20 years. His time at University College Cork has provided many opportunities to help communicate patterns and processes in ecology using this technology. He obtained his Master's degree in Zoology from UCC in 2006. His expertise is more related to freshwater systems, and in this field he has authored scientific papers and reports associated with the UN's Sustainable Development Goals.

Patricia Winters grew up on Clare Island, where she developed an interest in the native flora and fauna from a young age, ultimately leading to a master's degree in Environmental Science from Trinity College Dublin. A six-week bird survey of Clare Island was undertaken as part her final thesis, but she remains a keen observer of the island wildlife on her frequent visits. She currently works as an environmental officer with Mayo County Council and worked in similar roles in Dublin for over ten years.

NEW SURVEY OF CLARE ISLAND

New Survey of Clare Island Executive Committee 2020

Contributors to previous volumes in the New Survey of Clare Island series

Volume 1: History and Cultural Landscape: Timothy Collins, Críostóir Mac Cárthaigh (ed.), Nollaig Ó Muraíle, Kevin Whelan (ed.)

Volume 2: Geology: Peter Coxon, David Evans, John R. Graham (ed.), Kenneth T. Higgs, W.E.A. Phillips, C.J. Stillman, B.G.J. Upton, Michael Williams

Volume 3: Marine Intertidal Ecology: Michelle Cronin, Thomas Cross, Robert Cussen, Jane Delany, Michael Guiry, Louise Harrington, Christine Maggs, David McGrath, Alan Myers (ed.), Julia Nunn, Sandy O'Driscoll, Ruth O'Riordan, Anne Marie Power, Fabio Rindi

Volume 4: The Abbey: Ann Buckley, Ian Cantwell, Fergus Gillespie, Paul Gosling (ed.), Conleth Manning (ed.), Karena Morton, Micheál Ó Comáin, Christoph Oldenbourg, Roger Stalley, John Waddell (ed.)

Volume 5: Archaeology: Markus Casey, Michelle Comber, Paul Gosling (ed.), Maureen McCorry, Conleth Manning (ed.), Karen Molloy, Sharon Nestor, Adrian Phillips, Mike Williams, Paula King, John Waddell (ed.)

Volume 6: The Freshwater and Terrestrial Algae: Jenny Bryant, Michael Guiry (ed.), David John (ed.), T.K. McCarthy (ed.), Fabio Rindi (ed.), Patricia Sims, Brian Whitton, David Williamson

Volume 7: Plants and Fungi: Donal Synnott (ed.), Pete Coxon, Ryan Corcoran, Paul Gibson, Stephen McCarron, Timothy Ryle, Matthew Jebb, M.R.D. Seaward, D.H.S. Richardson and David Mitchel

Volume 8: Soils and Soil Associations: Wies Vullings, James F. Collins, George Smillie

Volume 9: Birds: R.H. Coombes, R.E. Cussen, R.E. Hutchinson, U. Keating, T.C. Kelly (ed.), E. McGreal, P. Smiddy, S. Warner, P. Winters

Clare Island, John Feehan

Forthcoming in the New Survey of Clare Island series

Volume 10: Zoology: R. Anderson, T. Bolger, K.G.M. Bond, J. Breen (ed.), M. Cawley, K. Creed, E. de Eyto, E. DeMilio, M. Gormally, C. Lawton, M. Cawley, M. Gormally, T.C. Kelly, E.S. Lenihan (ed.), N.J. Marley, K. McAney, G. McCall, T.K. McCarthy (ed.), S. McCormack, R. Mc Donnell, R. Møbjerg Kristensen, D.A. Murray, M. Nolan, A. O'Grady, R. Paxton, J. Petter Nilssen, A. Trowanska-Olichwer

PREFACE

The first Clare Island survey, the brainchild of Robert Lloyd Praeger, has received international acclaim (e.g. Berry, R.J. 2009 *The natural history of islands*. London. Collins; Feehan, J. 2019 *Clare Island*. Dublin. Royal Irish Academy). The 'Birds' paper included in that survey was written by Richard J. Ussher and was based on fieldwork conducted on the island between 1909 and 1911. Ussher's 'Aves' paper, however, also summarised details of the avifauna of a wider area in the west of Ireland—mainly the counties of Galway and Mayo—a theme that was revisited several times by the late Major Robert F. Ruttledge.

The current 'Birds' volume focuses exclusively on Clare Island and applies modern methods of census, particularly of seabirds and land birds. It develops a more detailed picture of the numbers of birds on the island than was possible at the time of the pioneer surveys in the 1909–11 interval. The fieldwork for this volume commenced broadly speaking in 1999 and continued intermittently until 2013. However, detailed seabird census data from 2015 to 2017 are also included.

Much work remains to be done however, and it is hoped that this volume will stimulate further research into the birds of Clare Island. The other volumes in the series including, in particular, those dealing with Marine Intertidal Ecology, Plants and Fungi, Soils and Soil Associations, Zoology as well as, of course, Geology and Archaeology, provide a uniquely detailed template in which to conduct *inter alia* habitat-specific avifaunal surveys, predator–prey relationships, population biology studies and competition between the different island land bird breeding species. Finally, genetic studies of some of the island's more sedentary species, especially perhaps the rock pipit *Anthus petrosus* (Montagu), might prove to be very informative and provide some definitive answers to the ambitions and questions addressed in the first survey.

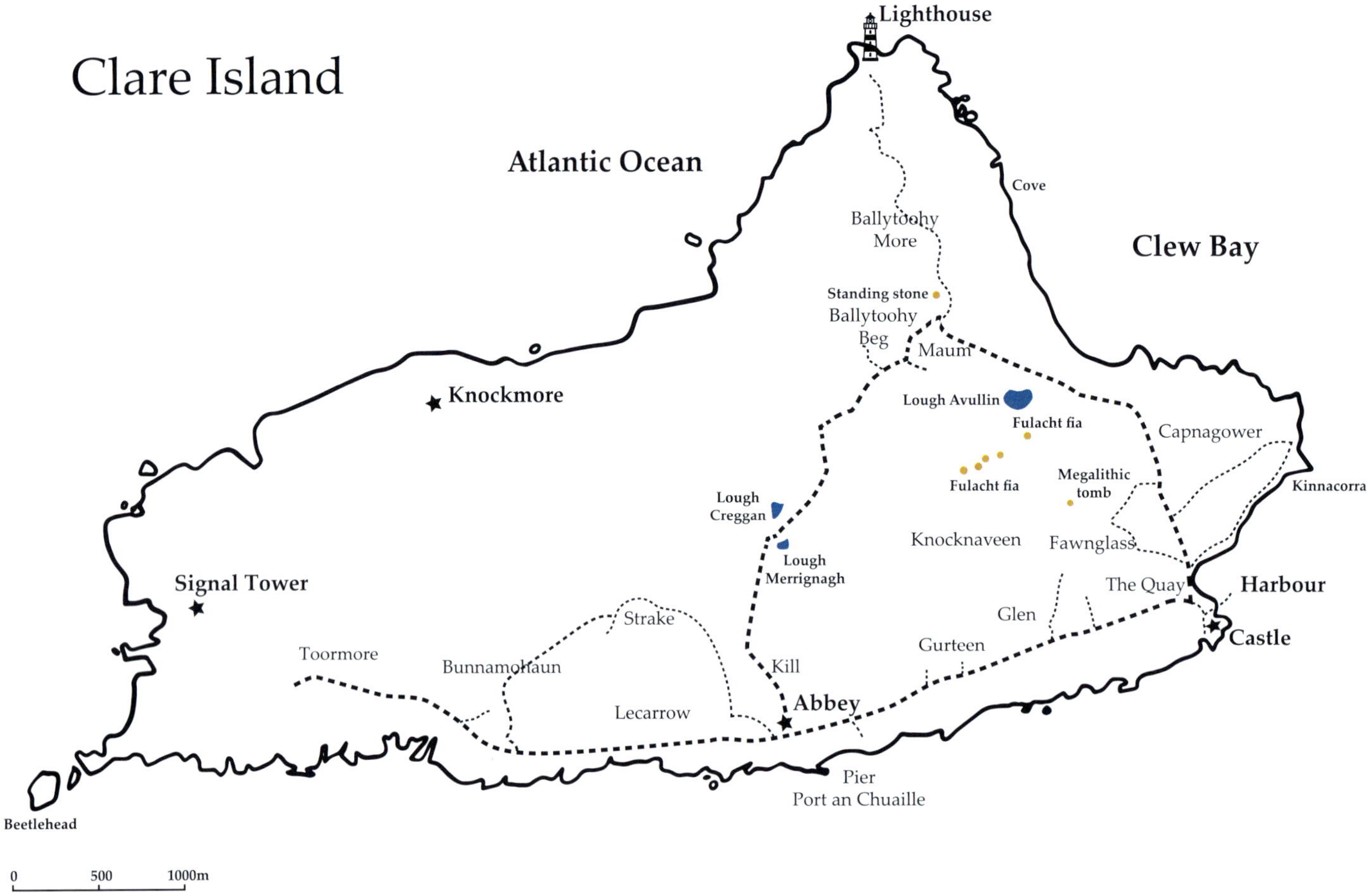
Clare Island
Lighthouse
Atlantic Ocean
Cove
Ballytoohy
More
Clew Bay
Standing stone
Ballytoohy
Beg
Maum
Lough Avullin
Knockmore
Fulacht fia
Capnagower
Megalithic
tomb
Fulacht fia
Kinnacorra
Lough
Creggan
Knocknaveen
Fawnglass
Lough
Merrignagh
Signal Tower
The Quay
Harbour
Glen
Strake
Castle
Gurteen
Toormore
Bunnamohaun
Kill
Abbey
Lecarrow
Pier
Port an Chuaille
Beetlehead
0
500
1000m

ACKNOWLEDGEMENTS

It is a pleasure to thank the people of Clare Island for the warmth of their welcome, patience with the activities of the surveyors, and for numerous and varied helpful discussions and advice.

We are deeply indebted to Professor Martin Steer for his expert advice and for the diplomacy and patience which he brought to bear to ensure the completion of this volume. We would also like to thank the anonymous referees for their incisive and constructive comments on the various manuscripts.

We acknowledge the expert assistance of Jonathan Dykes in the preparation and publication of this volume.

We are especially grateful to Richard T. Mills and Michael Ó Briain, who generously provided the bulk of the photographs used in this volume.

In compiling this volume, we have received data, constructive assistance and advice from a wide variety of colleagues, including Brian Caffrey in particular, whose contributions were crucial to the completion of Chapters 2, 4, 5, 6 and 7.

Professor Peter Coxon provided the approximate date of the origin of Clare Island as an island.

Others who assisted included Drs Paul M. Walsh, Olivia Crowe and Gavin Fennessy; Ricky Whelan (MSc) and Sam Bayley.

Many fieldworkers including the late Commandant Michael A. Hartnett, the late Dr Kieran McCarthy (NUIG), Professor John Feehan, and NPWS colleagues Ger O'Donnell, Aongus Ó Donail and Dermot Breen contributed to the data summarised in this volume. 'Michael Bob' ferried us around Clare Island in the unforgettable sea-based survey of June 1999.

Other colleagues, past and present, including Dr David Cabot, Dr Tim Sharrock and co-workers on Cape Clear Island, the late Major Robert Ruttledge and the late Clive D. Hutchinson contributed to our knowledge of the birds on Ireland's offshore islands.

Rachel E. Hutchinson, as well as participating in many of the fieldwork expeditions between 2003 and 2013, undertook the initial proofreading of the entire manuscript. Patrick Smiddy (MSc), in addition to his co-authoring contributions, also proofread all chapters and made numerous invaluable suggestions which greatly improved the manuscript; likewise, Usna Keating (MSc) made vital adjustments and corrections to the data and the analysis; similarly, Eoin McGreal updated the seabird data as well as adding considerably to the accuracy of the systematic list. Ms Patsy Winters (MSc) and Richard H. Coombes made substantial co-writing contributions as well as notable additions to the systematic list. Mr Stuart Warner (MSc), in addition to his edits, drew all the maps used in this volume.

We must also pay tribute to the pioneer ornithologists who made up the survey team of 1909–11 (listed in Ussher 1912), and to the participants of subsequent surveys including the late John Barlee and Major Robert Ruttledge (the breeding and winter atlas surveys of Sharrock (1976), Lack (1986), Gibbons *et al.* (1993) and Balmer *et al.* (2013)). Dr Clare Lloyd, the late Dr Tony Whilde, Mr Gordon D'Arcy, Mr Anthony McGeehan and those who contributed sightings to Mr Joe Doolan's www.irishbirding.com (which proved an invaluable resource in compiling the systematic list) all made an immense contribution to our knowledge of the avifauna of Clare Island.

It is hoped, therefore, that the current volume, which has a somewhat wider canvas than the 1909–11 study, is a fitting sequel to the pioneering investigations of the first survey and indeed to the surveyors themselves, as well as to all the participants in the subsequent expeditions.

Finally, while the elegance of expression, style and content may not reflect the typical 'islomania' (*sensu* Viney, M. 2014 *Irish Times* 31 May 2014) approach to the aesthetics and natural history of islands, it would be a mistake to think that the authors are not afflicted by this syndrome. Moreover, though referring to a greatly more rarefied level of thought and science, we cite the brilliant theoretical physicist Ludwig Boltzmann who observed that 'matters of elegance ought to be left to the tailor and cobbler' (quoted in Einstein, A. 1954 *Relativity*. London. Methuen).

Tom Kelly
February 2020

CHAPTER 1

INTRODUCTION

T.C. Kelly

Clare Island from the mainland. The lower peak Knocknaveen (223m) to the right and Cnoc Mór (463m) on the left can be clearly seen. Photo: T. Kelly

The birds, Class Aves, of Clare Island are the subject of this volume. In contrast to Ussher (1912), who summarised the findings of the first (1909–11) survey in one long paper, we have considered different aspects of the avifauna in separate chapters as follows: systematic list (Chapter 2), the seabirds of Clare Island (Chapter 3), the land birds of Clare Island (Chapter 4), waterbirds – ducks, rails and waders (Chapter 5), the land birds of the offshore islands of Ireland (Chapter 6), the equilibrium number of land birds of Clare Island (Chapter 7), the absence of the rook *Corvus frugilegus* Linnaeus from Clare Island (Chapter 8) and finally, the summary.

Clare Island (1,598ha), 5km (*c.* 3miles) at its closest point to the coast of County Mayo, was formed about 8,000 years before present (yBP). Rising sea levels, resulting from the melting of the massive ice sheets that formed during the last glacial maximum (LGM), 'drowned' the pre-existing coast and eventually separated the landmass that is now Clare Island from the adjacent mainland. At the peak of the LGM (*c.* 21,000 yBP) sea levels had reduced by 130 to 120m (Bintanja *et al.* 2005; Whittaker and Fernandez-Palacios 2007; Fernandez-Palacios *et al.* 2015). 'The rise of the late glacial minimum was not achieved overnight, nor was it a steady or uniform pattern' (Whittaker and Fernandez-Palacios 2007).

Therefore, despite the slow rise of the scends and gurges[1] to their present level (the same as when the island was first formed) the pre-existing fauna and flora on Clare Island would have become marooned. Of course, the pre-existing fauna and flora was itself a non-uniform assemblage, ranging from that which colonised the island in the immediate wake of the retreating glaciers with their periglacial climate envelopes, to the biota that became established as the climate ameliorated and was, presumably, very similar to that found on the island today (e.g. Pilcher and Hall 2001; Coxon 2008).

Isolation *per se* would have had a minimal impact on birds (Ussher 1912). Through the power of flight, which they share with bats (Mammalia: Chiroptera) and some insects, birds would have been highly mobile and able to move freely between the emerging island and the adjacent 'mainland' (Ussher 1912). Indeed, almost all offshore islands >500km^2 are host to many of the widely distributed breeding land birds in Britain and Ireland (see Balmer *et al.* 2013). For example, the blackbird *Turdus merula* Linnaeus is widely distributed from the Scilly Isles to the Shetland Isles and the dunnock *Prunella modularis* (Linnaeus) is distributed up to the Outer Hebrides and the Orkney Islands but not the Shetland group (Balmer *et al.* 2013). These distributions are, however, highly anomalous and difficult to explain in view of the analysis of Paradis *et al.* (1998) of the distances moved by birds that have been ringed in Britain and Ireland. Thus, breeding dunnocks, on (arithmetic) average, only move a distance of 1.4km from where they were ringed (mean natal movement 2.1km), while breeding blackbirds move an average of 3.2km (natal 3.3km) (see Appendix 1 and 2 in Paradis *et al.* 1998; see also Wernham *et al.* 2002). Clearly, these distances are too short to explain how either species was able to extend their ranges out to the offshore islands. So how were these islands colonised by these species? Notwithstanding the well-known critique of ringing as an unbiased estimator of dispersal distances (e.g. Koenig *et al.* 1996; Clobert *et al.* 2001; Wernham *et al.* 2002), part of the answer may lie in the 'long-tailed distribution' of dispersal distances detected by Paradis and co-workers (See Figs. 1 and 2 in Paradis *et al.* 1998). These data, for example, in the case of the blackbird, show that a small proportion of individuals (of unknown sex) in both the breeding and natal samples travelled much greater distances than the mean in the breeding sample (>300km) and in the natal sample (>350km). Such long-distance dispersers could be the individuals that establish founding populations on offshore islands. This dispersal polymorphism is well known in birds (e.g. Greenwood 1980; Clobert *et al.* 2009; Elliot and Cornell 2012) and *inter alia* is probably associated with 'personality' traits (see especially Canestrelli *et al.* 2015; Mettke-Hofmann 2017) as well as being gender and genetically determined (Saastamoinen *et al.* 2018).

However, environmental perturbations also play a role, particularly during intervals when ice is retreating; Udvardy and Papp (1969) review data published by the late Baron Lars Von Haartman on the pied flycatcher *Ficedula hypoleuca* (Pallas) which, in Northern Scandinavia is 'nomadizing and nonphilopatric', whereas in Central Europe the species is markedly philopatric, defined as the 'tendency or instinct to consistently return to their birthplaces or previous breeding grounds even after prolonged and regular juvenile wanderings or annual migratory flights'. Thus, it appears that as ice retreats, the 'normal' movement ecology is replaced by non-philopatric individuals whose dispersal behaviour enables a rapid range expansion into the areas vacated by the glaciers.

On Clare Island, just as with the other faunal and floral assemblages that formed as the ice sheets retreated, the pioneering avian community, which were adapted to periglacial environments (i.e. an Arctic/Boreal avifauna), would have been replaced by a group of species not greatly dissimilar to that found in Ireland today.

The first wave of postglacial colonising species may have included the partial migrants such as Lapland bunting *Calcarius lapponicus* (Linnaeus) and snow bunting *Plectrophenax nivalis* (Linnaeus), as well as the more sedentary forms including the rock ptarmigan *Lagopus muta* (Montin) and

[1] Arriving, and withdrawing waves, respectively.

gyr falcon *Falco rusticolus* Linnaeus. Migrants would have included many wildfowl and wading bird species including the brent *Branta bernicla* (Linnaeus) and greater white-fronted geese *Anser albifrons* (Scopoli), Eurasian whimbrel *Numenius phaeopus* (Linnaeus), turnstone *Arenaria interpres* (Linnaeus) and sanderling *Calidris alba* (Pallas) (e.g. Ratcliffe 1990; Kelly 2008; Kelly *et al.* 2014; Smiddy 2014). The northern wheatear *Oenanthe oenanthe* (Linnaeus), and particularly the Greenland race *O.o. leucorhoa* (J.F. Gmelin), which breeds in the Nearctic around eastern Canada and Greenland but migrates through Europe to Africa (e.g. Bairlein *et al.* 2012), would have been an early migrant passerine as the Holocene began. Of course, there was an ebb and flow of the periglacial adapted biota and those capable of optimising their life histories under more temperate conditions (see Coxon 2008 for details of the Nahanagan stadial which immediately preceded the commencement of the Holocene).

It appears that in the three millennia that followed the formation of Clare Island, mature woodland developed, including *Pinus*, *Quercus*, *Fraxinus* and *Ulmus* (Coxon *et al.* 2013). Thus, it seems reasonable to presume that a structured vegetation including tree, shrub and ground layers emerged over this interval. Such 'foliage height diversity' would, in turn, have enabled the eventual build-up of a diverse woodland bird community (Nairn and O'Halloran 2012; Fuller 2012).

Neolithic man arrived about 4,000yBP (e.g. Gosling 2007; Coxon *et al.* 2013; see also Mallory 2013), marking the beginning of the Anthropocene on Clare Island (e.g. Helmus *et al.* 2014; Corlett 2015). Over the succeeding 3,500 years or so, the woodland element was gradually removed and about 400 years ago the 'modern' agricultural landscape, including the *Calluna* formation (Doyle and Foss 1986; Ryle 2013), became established (Coxon *et al.* 2013). More recently (i.e. within living memory), the *Calluna* formation disappeared due, apparently, to intensive sheep grazing of the commonage (Doyle and Foss 1986; Ryle 2013; see also Feehan 2019).

The woodland bird community on Clare Island would probably have become mostly extinct as its habitat gradually disappeared over the millennia, and the more recent removal of the *Calluna* formation would have led to the demise of the red grouse *Lagopus lagopus* (Linnaeus) and possibly other species as well (e.g. the common whitethroat *Sylvia communis* Latham). However, agricultural activity and the expansion of grasslands would also have created opportunities for, *inter alia*, seed eating species and ground nesting forms such as the skylark *Alauda arvensis* Linnaeus and meadow pipit *Anthus pratensis* (Linnaeus) and migrants such as the northern wheatear and corncrake *Crex crex* (Linnaeus), and the extraction of peat would have created opportunities for wetland species (Coxon 2001; Vullings *et al.* 2013; Ryle 2013).

Well known synanthropic species such as the house sparrow *Passer domesticus* (Linnaeus), European robin *Erithacus rubecula* (Linnaeus), song thrush *Turdus philomelos* C.L. Brehm and perhaps the barn swallow *Hirundo rustica* Linnaeus would have also colonised the island (see especially O'Halloran *et al.* 2012). However, many of the above-mentioned species would have been vulnerable to changes in agricultural and land use practices (e.g. seed eaters such as the yellowhammer *Emberiza citrinella* Linnaeus) and to climate change.

In addition, the famines of the early to mid-nineteenth century would also have had an impact on the ecology of the inhabited offshore islands. For example, according to the letters of Henry Stratford Persse, the 1822 famine caused the departure of all islanders from Inisheer (Pethica and Roy 1998) and the population of Clare Island fell sharply after the Great Famine; substantial changes in the island's agricultural practices were to follow, with a major shift from arable land use to the production of grazing stock, especially sheep (Dwyer 1963; Doyle and Foss 1986). The abandonment of land, and the peoples who occupied that land, is well known to cause the departure of synanthropic bird species (e.g. Alexander 1954; Munns 1956). Although implied in a comment by Ó Fiannachta (1983) that birds effectively take over and make it agreeable for the 'bird watcher' when people leave a previously inhabited island, this is a portrayal that is not borne out by the facts.

The perturbations to the terrestrial ecosystem of Clare Island following the arrival of the Neolithic people some 4,000 years ago cannot, of course, be described in detail, but we can assume that the land bird avifauna on the island was significantly influenced by the recent island-wide changes in farming practices, the general impacts of the Anthropocene, and the pattern of colonisation of the island of Ireland by bird species mostly from Britain and the continent (Smiddy 2014).

This raises a question: how ancient is the land bird community on Clare Island? Or, to put it another way, what component of bird species within the assemblage are likely to have had a continuous presence on the island since the time of its formation some 8,000yBP and probably in the interval before it became finally separated?

The first survey 1909 to 1911

Island faunas and floras had been investigated since the late eighteenth century (Berry 2009); most of the early surveys were focused on compiling inventories, but following the work of Darwin (e.g. 1859) and Wallace (e.g. 1880), an evolutionary perspective was developed in which islands figured prominently (e.g. Losos and Ricklefs 2009). Praeger (1915) mentions Darwin and Wallace among others but only deals very briefly and somewhat tentatively with the potential for the Clare Island survey to reveal the evolutionary effects of isolation: 'and in islands which may have joined with adjoining areas in comparatively recent times we may hope to find some beginnings of that differentiation and specialization which is so remarkable a feature of the natural history of the oceanic islands'. However, Barrett-Hamilton (1915) is more assured, stating—in connection with the pygmy shrew *Sorex minutus* Linnaeus—'but had it survived the Glacial Period in Clare Island, such long isolation in the narrow confines of a small island might have been expected to have caused local differentiation of at least sub-specific value'.

Ussher (1912) stated that 'problems of isolation presented by the more sedentary organisms do not arise' in the case of flying animals such as birds. In relation to the land birds on Clare Island, he concluded that 'their characteristics are those of part of the whole of Ireland, and not the island, which is but a point in their range'. This conclusion seems to have followed the detailed examination of the approximately 82 skins collected by Barrington and Patten. However, these skins were collected presumably with the expectation of finding 'some beginnings of differentiation' that were so spectacular on truly oceanic (i.e. remote) islands of volcanic origin (e.g. Darwin 1859; Wallace 1880). Additionally, it may be relevant that in 1910, as the first survey of Clare Island was underway, three new subspecies of birds on the island of Ireland were added by taxonomists to its avifauna (Kennedy 1957; see also Mallet 2008). Finally, it is also possible that C.J. Patten (1870–1945) the Dublin-born and Trinity College-educated academic biologist who became professor of anatomy in the University of Sheffield (Heery 2018) may have been the 'theoretician' on the possibilities of 'differentiation' of Clare Island's birds.

It is interesting that while the authors (cited above) of the various papers arising from the findings of the first survey set out to find evidence of new taxa on the 'fragment island' (Warren *et al.* 2015) that is Clare Island, they do not quote or cite any authority in support of this hypothesis. Nevertheless, it seems that there was widespread acceptance of the origination of new species due to isolation (and presumably natural selection) by the various participants in the biological part of the first survey. However, the direct link between this hypothesis and the works of Darwin and Wallace is not easy to establish. For example, Mallet (2005) quotes (before a robust refutation) Mayr (1942) who stated that Darwin's (1859) classic was 'misnamed' because it was 'not a treatise on the origin of species'. The same can possibly be said of Wallace's (1880) book on island life (but see Mallet 2008; see also Quammen 1996).

The first Clare Island survey was undertaken before the 'modern synthesis' (see Ridley 2004) and the emergence of Neo-Darwinism. However, the idea of the formation of a new species due to isolation was a precursor of Mayr's biological species concept, which followed Dobzhansky's genetic interpretation of geographical and reproductive isolation (such as would occur on islands) (for a detailed review see Ayala and Fitch 1997) as the major mechanism in allopatric speciation (*sensu* Mayr 1991). The biological species concept as an explanatory model of how new species arise has been adopted by most, but not all, evolutionary biologists (e.g. Uy *et al.* 2018). However, Mayr (1963) also proposed the 'founder effect', sometimes also referred to as the peripheral isolate model, as a major mechanism in allopatric speciation, particularly on islands, but this idea has fallen into 'disfavour' (Matute 2013; see also, for example, Moya *et al.* 1995; Coyne and Orr 2004).

Future investigations into evolutionary aspects of the avifauna of Clare Island will not rely wholly on the archived 'skins' as modern techniques of mist-netting and the removal of miniscule amounts of tissue for DNA studies will not harm birds. The taking of skins was widely practiced in the nineteenth and early twentieth centuries: 'this day I shot a condor' (Darwin 1845; see also Quammen 1996). Too much time has passed to estimate the impact (in terms of possible additive, that is, non-compensatory mortality) of the removal of relatively large numbers of animals—including from Clare Island—on the demographic viability of the residual populations.

At least four species might be worth detailed investigation in the future, based on an integrated taxonomic study (i.e. one involving mitochondrial and nuclear DNA), the physical phenotype including biometrics, and behavioural aspects including vocalisations. The rock pipit *Anthus petrosus* (Montagu) is effectively sedentary in Ireland (Hutchinson 1989; Wernham *et al.* 2002) and has probably lived continuously on Clare Island since its formation. The dunnock, wren *Troglodytes troglodytes* (Linnaeus) and the reed bunting *Emberiza schoeniclus* (Linnaeus) are similarly sedentary (Wernham *et al.* 2002) and possibly may also have survived continuously on Clare Island over the 8,000-year interval since its formation.

The theory of island biogeography

Island faunas and floras became a major focus of science once again following the publication of MacArthur and Wilson's (1967) monograph '*The theory of island biogeography*' and an earlier paper (1963) entitled 'An equilibrium theory of insular zoogeography'. These publications, though not uncontroversial (e.g. Quammen 1996; Williamson 1989), have generated an enormous amount of research into all aspects of island ecology and have been applied to, for example, nature reserve selection (e.g. Whittaker and Fernandez-Palacios 2007; Losos and Ricklefs 2010).

The detail on plant and invertebrate diversity as well as birds and mammals presented in the first Clare Island survey could be used to test some of the island biogeography theory 'rules', particularly those concerned with the relationships between species richness and island area, and distance from the mainland. In general, species richness is positively correlated with island area and inversely with distance from the mainland.

More recently, island biogeography theory (henceforward IBT) has been broadened to incorporate the impact of glaciation, particularly the Last Glacial Maximum (LGM) on the grounds *inter alia* that fluctuating sea levels would have altered the area of islands and the distances between islands and the mainland (e.g. Whittaker *et al.* 2008, 2010, 2017; Fernandez-Palacios *et al.* 2015; Weigelt *et al.* 2016; Fernandez-Palacios 2016). These latest findings are relevant to the potential impact of climate change-related rising sea levels on island areas, and, therefore, their species richness, as well as to questions relating to the origin of the post-glacial fauna and flora of Ireland (e.g. Beirne 1952; Sleeman *et al.* 1986; Davenport *et al.* 2008; Sleeman *et al.* 2014) and, more specifically, the birds (McCarthy 1986; Hutchinson 1989; Kelly 2008; Kelly *et al.* 2014; Smiddy 2014).

Bird migration on Clare Island

Apart from Barrington (1900) there has been no systematic study of bird migration on Clare Island. Most land bird migration takes place at night (Alerstam 1990). In this regard, Barrington's (1900) study based on the observations of light keepers and the collection of cadavers that had struck the lights, was of outstanding interest. However, while the study did include Clare Island, very little data seems to have been collected at the island's lighthouse. Part of the explanation may be that the candle power of the light was quite low, and the generation of the light was by reflection rather than refraction (Barrington 1900) though other authorities (e.g. Ruttledge 1950) argued that the lighthouse was at too great an elevation (*c.* 400 feet) to attract large numbers of migrants. However, Barrington (1900) was also critical of the light keepers themselves describing their performance in submitting completed schedules and bird remains as 'fair' in 1894 and 'poor' in 1895.

Interestingly, Ruttledge (1947) records a 'remarkably large' gathering of between 200 and 300 northern wheatears 'around the light for two hours from 2 a.m. on 29 March 1946', and 25 to 30 were at the light on the night of 30 March, all of which were observed by Mr P.A. McMahon. The latter also saw a great skua *Stercorarius skua* (Brünnich) flying in from the sea and attacking a European herring gull *Larus argentatus* Pontoppidan on 2 September 1947. The lighthouse, which was built in 1806, was decommissioned in 1965.

The diurnal passage migration of seabirds has not been addressed in any detail, but casual observations suggest that the low-lying point at Kinacorra and, at the opposite end of the island, the Signal Tower, might be worth monitoring when the winds from both south westerly and north westerly directions increase in force (T.C. Kelly, unpublished data).

Conservation and the avifauna of Clare Island

Part of Clare Island is designated as a Special Protection Area (SPA) for listed seabirds and the chough *Pyrrhocorax pyrrhocorax* (Linnaeus) (see http://www.irishstatutebook.ie/eli/2010/si/273/made/en/pdf).

Several red-listed species (Colhoun and Cummins 2013) including the corncrake, twite *Carduelis flavirostris* (Linnaeus) and yellowhammer have all become extinct relatively recently on Clare Island, but two others, the northern lapwing *Vanellus vanellus* (Linnaeus) and European herring gull, remain (see Chapter 2). The population of the European herring gull is increasing (Chapter 3) but that of the northern lapwing remains small and is probably vulnerable to extinction (Chapter 5). The creation of more permanent wetland habitat, which has decreased considerably since Praeger's (1911) survey (Vullings *et al.* 2013), might stimulate an increase in the number of northern lapwing and other waterbird species.

Future genetic studies on Clare Island and indeed elsewhere in Ireland may be revealing and germane from both a conservation and IBT perspective; some species with apparent phenotypic similarity show remarkable genetic differentiation (e.g. Venkatraman *et al.* 2019) whereas others show a high level of phenotypic variation but relatively little underlying genetic differentiation (e.g. pied wagtail *Motacilla alba* Linnaeus) (Semenov *et al.* 2018; see also Coates *et al.* 2018).

Clare Island, therefore, is something of an insular laboratory with specimen and skin collections dating back to 1909. Hopefully, the island's flora and fauna—including birds—will be revisited by a programme of integrated taxonomy. Recent advances in genomics underline the views of Riddle and Hafner (1999) who suggest that biogeographers and conservation ecologists should consider the 'evolutionary continuum to species' (Coates *et al.* 2018), including 'evolutionarily significant units' (ESU) as well as, for example, cryptic species (Fišer *et al.* 2018) when estimating biodiversity.

Endnote

Richard Ussher died in 1913, only one year after the publication of the Clare Island paper, and R.M. Barrington died in 1915. Both made remarkable contributions to the documentation and understanding of different aspects of the avifauna of Ireland (Hutchinson 1997).

REFERENCES

Alerstam, T. 1990 *Bird Migration*. Cambridge. Cambridge University Press.

Alexander, S.M.D. 1954 The birds of the Blasket Islands, with special reference to the Great Blasket Island, Inishvickillaun and Illaunboy and some notes on the adjacent mainland. *Bird Study* **1**, 148–68.

Ayala, F.J. and Fitch, W.M. 1997 Genetics and the origin of species: an introduction. *Proceedings of the National Academy of Sciences* **94**, 7691–7.

Bairlein, F., Norris, D.R., Nagel, R., Bulte, M., Voight, C.C., Fox, J.W., Hussell, D.J. and Smaljohann, H. 2012 Cross-hemisphere migration of a 25g songbird. *Biological Letters* **8**, 505–7.

Balmer, D.E., Gillings, S., Caffrey, B.J., Swann, R.L., Downie, I.S. and Fuller, R.J. 2013 *Bird Atlas 2007–2011: the breeding and wintering birds of Britain and Ireland*. Thetford. BTO Books.

Barrett-Hamilton, G.E.H. 1915 Mammalia. A biological survey of Clare Island in the county of Mayo, Ireland and adjoining district. *Proceedings of the Royal Irish Academy* **31,** Sections 1–3, Part 17.

Barrington, R.M. 1900 *Migration of birds as observed at Irish Lighthouses and Lightships*. Dublin and London. Ponsonby and Porter.

Beirne, B.P. 1952 *The origin and history of the British fauna*. London. Methuen.

Berry, R.J. 2009 *The Natural History of Islands*. London. Collins.

Bintanja, R., van de Wal, R.S. and Oerlaemans, J. 2005 Modelled atmospheric temperatures and global sea levels over the past million years. *Nature* **437**, 125–8.

Canestrelli, C., Bisconti, R. and Carere, C. 2015 Bolder takes all? Behavioural dimension to biogeography. *Trends in Ecology and Evolution* **31**, 35–43

Clobert, J., Danchin, E., Dhondt, A.A. and Nicols, J.D. 2001 *Dispersal*. Oxford. Oxford University Press.

Clobert, J., Le Gaillard, J.F., Cote, J., Meylan, S. and Massot, M. 2009 Informed dispersal, heterogeneity in animal dispersal syndromes and the dynamics of spatially structured populations. *Ecology Letters* **12**, 197–209.

Coates, D.J., Byrne, M. and Moritz, C. 2018 Genetic diversity and conservation units: dealing with

the species – population continuum in the age of genomics. *Frontiers in Ecology and Evolution* **6**, Article 165.
Colhoun, K. and Cummins, S. 2013 Birds of Conservation Concern in Ireland. *Irish Birds* **9**, 523–44.
Corlett, R.T. 2015 The Anthropocene concept in ecology and conservation. *Trends in Ecology and Evolution* **30**, 36–41.
Coxon, P. 2001 The quaternary history of Clare Island. In J.R. Graham (ed.) *New Survey of Clare Island* Volume **2**, 87–112. Dublin. Royal Irish Academy.
Coxon, P. 2008 Landscapes and environments in the late glacial-interglacial transition: a time of amazingly rapid change in Ireland. In J.L. Davenport, D.P. Sleeman and P.C. Woodman (eds) *Mind the gap: postglacial colonization of Ireland*. 45–61. Belfast. Irish Naturalists' Journal.
Coxon, P., Corcoran, R., Gibson, P. and McCarron, S. 2013 A Holocene pollen diagram from Lough Avullin, Clare Island, western Ireland. In D. Synnott (ed.) *New Survey of Clare Island*. Volume **7**, 1-–26. Dublin. Royal Irish Academy.
Coyne, J.A. and Orr, A. 2004 *Speciation*. Massachusetts, Sinaur Associates.
Darwin, C. 1845 *The Voyage of the Beagle*. London. John Murray.
Darwin, C. 1859 *On the origin of species by means of natural selection, or the preservation of favoured races in the struggle for life*. London. John Murray.
Davenport, J.L., Sleeman, D.P. and Woodman, P.C. (eds) 2008 *Mind the gap: postglacial colonization of Ireland*. Special Supplement 2008. Belfast. Irish Naturalists' Journal.
Doyle, G.J. and Foss, P.J. 1986 A resurvey of the Clare Island flora. *Irish Naturalists' Journal* **22**, 85–9.
Dwyer, D.J. 1963 Farming an Atlantic Outpost: Clare Island, Co. Mayo. *Geography* **48**, 255–67.
Elliot, E.C. and Cornell, J. 2012 Dispersal polymorphism and the speed of biological invasions. *PLoS One* **7**, 00404496.
Feehan, J. 2019 *Clare Island*. Dublin. Royal Irish Academy.
Fernandez-Palacios, J.M. 2016 Shaped by sea level shifts. *Nature* **532**, 42–3.
Fernandez-Palacios, J.M., Rijsdijk, K.F., Norder, S.J., Otto, R., de Nascimento, L., Fernandez-Lugo, S., Tjorve, E. and Whiitaker, R.J. 2015 Towards a glacial-sensitive model of island biogeography. *Global Ecology and Biogeography* **25**, 817–30.
Fǐser, C., Robinson, C.T. and Malard, F. 2018 Cryptic species as a window into the paradigm shift of the species concept. *Molecular Ecology* **27**, 613–35.
Fuller, R.J. (ed.) 2012 *Birds and Habitat. Relationships in changing landscapes*. Cambridge. Cambridge University Press.
Gosling, P. 2007 The human settlement history of Clare Island. In P. Gosling, C. Manning and J. Waddell (eds) *New Survey of Clare Island*. Volume **5**, 29–68. Dublin. Royal Irish Academy.
Greenwood, P. 1980 Mating systems, philopatry and dispersal in birds and mammals. *Animal Behaviour* **28**, 1140–62.
Heery, S. 2018 Charles Joseph Patten: professor and Irish ornithologist. *Irish Birds* **11**, 101–5.
Helmus, M.R., Mahler, L. and Losos, J.B. 2014 Island biogeography of the Anthropocene. *Nature* **513**, 543–6.
Hutchinson, C.D. 1989 *Birds in Ireland*. Calton. Poyser.
Hutchinson, C.D. 1997 Bird study in Ireland. In J.W. Foster (ed.) *Nature in Ireland – a scientific and cultural history*. 262–82. Dublin. Lilliput Press.
Kelly, T.C. 2008 The origin of the avifauna of Ireland. In J.L. Davenport, D.P. Sleeman and P.C. Woodman (eds) *Mind the gap: postglacial colonization of Ireland*. 97–107. Belfast. Irish Naturalists' Journal.
Kelly, T.C., Smiddy, P. and Graham, C.T. 2014 Origins of the Irish avifauna – a perspective from Island Biography Theory. In D.P. Sleeman, J. Carlsson and J.E.L. Carlsson (eds) *Mind the gap II. New insights into the Irish postglacial*. 44–52. Belfast. Irish Naturalists' Journal.
Kennedy, P.G. 1957 Birds. In J. Meenan and D.A. Webb (eds) *A View of Ireland*. 62–8. British Association for the Advancement of Science. Dublin. Hely's Limited.
Koenig, W.D., Von Vuren, D. and Hooge, P.N. 1996 Detectability, philopatry and the distribution of dispersal distances in vertebrates. *Trends in Ecology and Evolution* **11**, 514–7.
Losos, J.B. and Ricklefs, R. 2009 Adaptation and diversification on islands. *Nature* **457**, 830–6.
Losos, J.B. and Ricklefs, R. (eds) 2010 *The Theory of Island Biogeography Revisited*. Princeton and Oxford. Princeton University Press.
MacArthur, R.H. and Wilson, E.O. 1963 An equilibrium theory of insular zoogeography. *Evolution* **17**, 373–87.
MacArthur, R.H. and Wilson, E.O. 1967 *The Theory of Island Biogeography*. Monographs in Population Biology, Volume 1. Princeton. Princeton University Press.
Mallet, J. 2005 Speciation in the 21st century. A review of "Speciation" by J.A. Coyne and H. Allen Orr. *Heredity* **95**, 105–9.
Mallet, J. 2008 A century of evolution: Ernst Mayr (1904–2005). Mayr's view of Darwin: was Darwin wrong about speciation? *Biological Journal of the Linnaean Society* **95**, 3–16.
Mallory, J.P. 2013 *The Origins of the Irish*. London. Thames and Hudson.
Matute, D.R. 2013 The role of founder effects on the evolution of reproductive isolation. *Journal of Evolutionary Biology* **26**, 2299–311.
Mayr, E. 1942 *Systematics and the Origin of Species*. New York. Columbia University Press.
Mayr, E. 1963 *Animal Species and Evolution*. Massachusetts. Harvard University Press.
Mayr, E. 1991 *One long argument – Charles Darwin and the genesis of modern evolutionary thought*. London. Penguin Press.
McCarthy, T.K. 1986 Biogeographical aspects of Ireland's invertebrate fauna. In D.P. Sleeman, R. Devoy and P.C. Woodman (eds) *Post Glacial Colonization of Ireland. Occasional Publication of the Irish Biogeographical Society* **1**: 67–81.
Mettke-Hofmann, C. 2017 Avian movements in a modern world: cognitive challenges. *Animal Cognition* **20**, 77–86.
Moya, A., Galiana, A. and Ayala, F.J. 1995 Founder effect speciation theory: failure of experimental corroboration. *Proceedings of the National Academy of Sciences* **92**, 3983–6.

Munns, D.J. 1956 Further notes on the birds of the Blasket Islands. *Bird Study* **3**, 248–50.

Nairn, R. and O'Halloran, J. (eds) 2012 *Bird Habitats in Ireland*. Cork. Collins Press.

Ó Fiannachta, P. 1983 An tOileánach: The Islandman by Tomas Ó Criomhthain. In P. Ó Conluain (ed.) *Islands and Authors. Pen pictures of life past and present on the islands of Ireland*. 44–58. Cork. Mercier Press.

O'Halloran, J., Nairn, R., Smiddy, P. and Crowe, O. 2012 Bird Habitats: a synthesis and future perspectives. In R. Nairn and J. O'Halloran (eds) *Bird Habitats in Ireland*. 247–68. Cork. Collins Press.

Paradis, E., Baillie, S.R., Sutherland, W.J. and Gregory, R.D. 1998 Patterns of natal and breeding dispersal in birds. *Journal of Animal Ecology* **67**, 518–36.

Pethica, J.L. and Roy, J.C. (eds) 1998 *'To the Land of the Free from this Island of Slaves'. Henry Stratford Persse's letters from Galway to America, 1821–1832*. Cork. Cork University Press.

Pilcher, J. and Hall, V. 2001 *Flora Hibernica*. Cork. Collins Press.

Praeger, R.L. 1911 Clare Island Survey, Part 10, Phanerogamia and Pteridophyta. *Proceedings of the Royal Irish Academy* **31**, 1–112.

Praeger, R.L. 1915 General introduction and narrative. *Proceedings of the Royal Irish Academy* **31**(1911–1915), Section 1, Part 1 (1915), 1–3.

Quammen, D. 1996 *The Song of the Dodo*. London. Hutchinson.

Ratcliffe, D. 1990 *Bird Life of Mountain and Woodland*. Cambridge. Cambridge University Press.

Riddle, B.R. and Hafner, D.J. 1999 Species as units of analysis in ecology and biogeography: time to take the blinkers off. *Global Ecology and Biogeography* **8**, 433–41.

Ridley, M. 2004 *Evolution*. Third edition. Oxford. Blackwell.

Ruttledge, R.F. 1947 Report on birds at Irish Light-stations 1946. *Irish Naturalists' Journal* **9**, 57–61.

Ruttledge, R.F. 1950 A list of the birds of the Counties of Galway and Mayo: showing their status and distribution. *Proceedings of the Royal Irish Academy* **52B**, 315–81.

Ryle, T. 2013 Vegetation-Environment interactions on Clare Island. In D. Synnott (ed.) *New Survey of Clare Island*. Volume **7**, 27–178. Dublin. Royal Irish Academy.

Saastamoinen, M., Bocedi, G., Cote, J., Legrand, O., Guillome, F., Wheat, C.W., Fronhofer, E.A., Garcia, C., Henly, R., Husby, A., Baguette, M., Bonte, D., Coulon, A., Kokko, H., Mathyssen, E., Niitepõld, K., Nonaka, E., Stevens, V.M., Travis, J.M.J., Donoghue, K., Bullock, J.M. and del Mar Delgado, M. 2018 Genetics of dispersal. *Biological Reviews* **93**, 574–99.

Semenov, G.A., Koblik, E.A., Red'Kin, Y.A. and Badyaev, A.V. 2018 Extensive phenotypic diversification coexists with little genetic divergence and lack of population structure in the white wagtail subspecies complex. *Journal of Evolutionary Biology* doi:10.1111/jeb.13305.

Sleeman, D.P., Carlsson, J. and Carlsson, J.E.L. (eds) 2014 *Mind the gap II: new insights into the Irish postglacial*. Belfast. Irish Naturalists' Journal.

Sleeman, D.P., Devoy, R. and Woodman, P.C. (eds) 1986 *Post Glacial Colonization of Ireland*. Occasional Publication of the Irish Biogeographical Society **1**.

Smiddy, P. 2014 Breeding birds in Ireland: success and failure among colonists. In D.P. Sleeman, J. Carlsson and J.E.L. Carlsson (eds) *Mind the gap II: new insights into the Irish postglacial*. 89–99. Belfast. Irish Naturalists' Journal.

Udvardy, M.D.F. and Papp, C.S. 1969 *Dynamic Zoogeography*. London. Von Nostrand Reinhold Company.

Ussher, R.J. 1912 A biological survey of Clare Island in the county of Mayo, Ireland, and adjoining district. *Proceedings of the Royal Irish Academy* **31**, Section 11, Part 20.

Uy, J.A.C., Irwin, D.E. and Webster, M.S. 2018 Behavioral isolation and incipient speciation in birds. *Annual Review of Ecology, Evolution and Systematics* **49**, 1–24.

Venkatraman, M.X., Deraad, D.A., Tsai, W.L.E., Zarza, E., Zellmer, A.J., Maley, J.M. and McCormack, J.E. 2019 Cloudy with a chance of speciation: integrative taxonomy reveals extraordinary divergence within a Mesoamerican cloud forest bird. *Biological Journal of the Linnaean Society* **126**, 1–15.

Vullings, W., Collins, J.F. and Smillie, G. 2013 Soils and soil associations. *New Survey of Clare Island*. Volume **8**, 1–176. Dublin. Royal Irish Academy.

Wallace, A.R. 1880 *Island Life: or the causes of insular faunas and floras, including a revision and attempted solution of the problem of geological climates*. London. Macmillan.

Warren, B.H., Simberloff, D., Ricklefs, R.E., Aguile, R., Condamine, F.L., Gravel, D., Morlon, H., Mouquet, N., Rosindell, J., Casquet, J., Cornault, J., Fernandez-Palacios, J.M., Hengl, T., Norder, S.J., Rijsdijk, K.F., Sanmartin, D., Triantis, K.A., Valente, L.M., Whittaker, R.J., Gillespie, R.G., Emerson, B.C. and Thebaud, C. 2015 Islands as model systems in ecology and evolution: prospects fifty years after MacArthur-Wilson. *Ecology Letters* **18**, 200–17.

Weigelt, P., Steinbauer, M.J., Cabral, J.S. and Kreft, H. 2016 Late Quaternary climate change shapes island biodiversity. *Nature* **532**, 99–102.

Wernham, C., Toms, M., Marchant, J., Clark, J., Siriwardina, G. and Baillie, S. (eds) 2002 *The Migration Atlas: movement of the birds of Britain and Ireland*. London. Poyser.

Whittaker, R.J. and Fernandez-Palacios, J.M. 2007 *Island Biogeography*. Second edition. Oxford. Oxford University Press.

Whittaker, R.J., Fernandez-Palacios, J.M., Matthews, T.J., Borregaard, M.K. and Triantis, K.A. 2017 Island biogeography: taking the long view of nature's laboratories. *Science* **357**, eeam8236 doi:10.1126/science.eeam8326.

Whittaker, R.J., Triantis, K.A. and Ladle, R.J. 2008 A general dynamic theory of oceanic island biogeography. *Journal of Biogeography* **36**, 977–94.

Whittaker, R.J., Triantis, K.A. and Ladle, R.J. 2010 A general dynamic theory of oceanic island biogeography: extending the MacArthur and Wilson theory to accommodate the rise and fall of volcanic islands. In J.B. Losos and R. Ricklefs (eds) *The Theory of Island Biogeography Revisited*. 88–115. Princeton. Princeton University Press.

Williamson, M. 1989 The MacArthur – Wilson theory today: true but trivial. *Journal of Biogeography* **16**, 3–4.

CHAPTER 2

SYSTEMATIC LIST OF THE BIRDS OF CLARE ISLAND

T.C. Kelly, R.H. Coombes, P. Winters, R.E. Cussen, E. McGreal, R.E. Hutchinson and P. Smiddy

ABSTRACT

This Systematic List of the birds of Clare Island contains an inventory of about 123 species (there are doubts that a few of the listed species have ever occurred on Clare Island), and this is compared with the findings of the first survey conducted during 1909–11. The list comprises records that span the period from 1887 to 2018. While some species have been added to the breeding community, e.g. willow warbler *Phylloscopus trochilus* (Linnaeus), goldcrest *Regulus regulus* (Linnaeus) and, most remarkably, the great skua *Stercocorarius skua* (Brünnich), others have disappeared including the corn bunting *Emberiza calandra* (Linnaeus), yellowhammer *Emberiza citrinella* Linnaeus, common cuckoo *Cuculus canorus* Linnaeus and corncrake *Crex crex* (Linnaeus). Many of these changes reflect parallel trends in the status of these species on the island of Ireland as a whole. The ring ouzel *Turdus torquatus* Linnaeus is formally removed from the inventory as there is insufficient proof that it has ever been identified on the island. Overall, the relatively short systematic list—compared to, for example, Cape Clear, Co. Cork (Sharrock 1973)—reflects the fact that the island is visited relatively infrequently by skilled field ornithologists, especially in the periods of peak migration in spring and autumn.

Introduction

A systematic list is one in which all the species that have been reported to occur at a given location are listed and each account is accompanied by, or annotated with, a brief set of notes. That is broadly the format followed by Ussher (1912) and many others (e.g. Sharrock 1973; Dymond 1980, 1991; Betts 1992; Lysaght 2002). However, in this volume, detailed accounts of the breeding seabirds (Chapter 3), breeding land birds (Chapter 4), breeding waterbirds (Chapter 5), breeding land birds on the offshore islands of Ireland (Chapter 6), the equilibrium number of species on Clare Island (Chapter 7), and the absence of the rook as a breeding species (Chapter 8) are presented, so the species that are focused upon in these chapters are accompanied by only brief notes in this inventory (scientific names of species mentioned are given in the systematic list).

In addition, the 'inventory' is 'systematic' because the order in which the species are listed and the families in which they are placed is based on a multifaceted determination of the evolutionary relationships between the taxa (e.g. Mayr 1999). However, the systematics of birds (Class Aves) have changed profoundly since Ussher's (1912) paper—the systematic names in this part are those in Howard Saunders' 'Manual' (1899). Saunders' Manual begins with the passerines and near passerines and leads on to the owls and raptors, pelecaniform species and so forth (Ussher 1912; see also Ruttledge 1950). Of course,

The great northern diver *Gavia immer* (Brünnich) is a regular winter visitor and can be seen from elevated ground above the harbour beach (pictured) into at least mid-April. Photo: T. Kelly

Ussher (1912) recognised that this system would be 'superseded' and so it was and indeed continues to be. For example, Hobbs (2018) has produced an excellent new list of 'Birds in Ireland' that is based on Gill and Donsker (2018), a new taxonomy in which the accipiters, i.e. the hawks, such as the sparrowhawk and its allied species, have been placed in a new Order Accipitriformes, separate from the Order Falconiformes (the falcons) which contains, for example, the familiar common kestrel and peregrine falcon, and within which they had been previously included (e.g. BWP*i* V2). However, a robust debate is ongoing on issues such as the 'species concept' within the voluminous literature on the systematics and evolution of the Class Aves (e.g. Gill 2014; Sangster 2014; Toews 2015). Barrowclough *et al.* (2016), in a study that combined morphological and molecular data, concluded that avian species diversity is underestimated by 'at least a factor of two'. Further, they suggest that the number of avian taxa may be as high as 18,043—compared to the currently accepted '9,159 biological (bird) species'; others put the current total number of bird species at 10,828 (Tietze 2018).

Part of the continuing modification of avian systematics has been due to the application of modern techniques of molecular genetics, especially mitochondrial DNA (mtDNA), nuclear DNA, and, more recently, whole genomes in the classification and phylogeny of birds (e.g. McKay and Zink 2010; Ellegren 2013; Prum *et al.* 2015; Craig *et al.* 2018; Sangster 2018). This 'integrative taxonomy' (Sangster 2018), which 'employs multiple types of evidence' including genomics, is likely to change existing systematics quite profoundly (see Prum *et al.* 2015, and references therein) resulting in a reinterpretation of phenomena such as avian phylogeography and biogeography (e.g. Newton 2003; Kelly *et al.* 2014; Smiddy 2014; Uva *et al.* 2018; Alstrom *et al.* 2018). Integrative taxonomy is also exerting an increasing influence on the higher-level phylogeny of Neornithine birds (Mayr 2011; see also Kapusta and Suh 2017).

As the integrative taxonomy approach, especially that which includes nuclear DNA (e.g. introns only) and whole genomes, is still in a relatively early stage of development, we have opted to follow Parkin and Knox (2010), which still commands wide support among avian systematists and taxonomists in Britain and Ireland; species and subspecies nomenclature also follows Parkin and Knox (2010).

It should also be noted that Clare Island does not have a long systematic list, i.e. one that includes many different species such as would be found, for example, at Cape Clear (Sharrock 1973) and Great Saltee Islands (Ruttledge 1963); some common species such as the mute swan *Cygnus olor* (J.F. Gmelin) and kingfisher *Alcedo atthis* (Linnaeus) have not been definitely recorded on the island. Clare Island does not appear to have been visited by many ornithologists other than during the breeding season (see for example Dempsey and O'Clery 2007), where most of the focus on the avifauna has been directed. The peak migration times of April and May and September and October have received much less attention. Barrington (1900) is the only available source of information on the migration of birds at Clare Island, although almost all the data were collected by the then resident lighthouse keepers in the late nineteenth century.

A note on the methods

The systematic list presented here summarises the published findings of Barrington (1900), Ussher (1912), Barlee and Ruttledge (1945), Lloyd (1984), D'Arcy (1992), Forsyth and Allen (1997), and Winters (2006) together with the unpublished observations of Robert Cussen (REC), Patricia Winters (PW), Richard H. Coombes (RHC), Eoin McGreal (EM) and T.C. Kelly (TCK). Other sightings have been reported on the internet, especially at (www.irishbirding.com) managed by Joe

Doolan, and this database has been accessed to obtain the latest information. The systematic list does not focus on a season or year; rather it sets out to summarise all information that is available on all species that have been seen on Clare Island.

Furthermore, in this account we have relied heavily on the various breeding and winter atlas studies that have been completed over the past 35 years. These, broadly speaking, follow a similar basic method in that all records refer to one, or both, of the two 10km squares (L68 and L78) that encompass Clare Island. L68 covers most of the island, whereas a relatively small strip along its eastern margin is contained within L78. Consequently, only a small number of tetrads within L78 refer to Clare Island; the remainder—the much larger part—includes the adjacent mainland.

Consequently, the maps in Sharrock (1976), Lack (1986), Gibbons *et al.* (1993) and Balmer *et al.* (2013) had to be checked against the original database to ensure that the putative record referred to Clare Island, rather than the mainland.

Finally, it should be remembered that a variety of different techniques were used to detect birds in the different atlas studies (for details consult the above listed references). Additional techniques, employed in our own studies, are specified in Chapters 3, 4, 5 and 7. It should be noted that these surveys were more focused in estimating the numbers of individuals belonging to the different species rather than searching for additional diversity within the island's avian community.

Note: nomenclature follows Parkin and Knox (2010) and Smiddy (2010).

SYSTEMATIC LIST

Family Anatidae

Greater White-fronted Goose *Anser albifrons* (Scopoli)

Greenland White-fronted Goose *Anser albifrons flavirostris* Delgaty and Scott

Rare straggler in the past; no recent records

There are no recent records on Clare Island but Ruttledge (1950) quotes P. McMahon who states that a few 'occasionally visit Clare Island'. However, Clare Island is not mentioned as a haunt in the comprehensive and detailed review of Ruttledge and Ogilvie (1979) or in Ruttledge (1994).

Greylag Goose *Anser anser* (Linnaeus)

Rare vagrant: one record

One was seen near the shore in Lecarrow on 15 December 2018 (PW). Greylag geese (feral population) are breeding on several islands to the south of Clare Island (EM) and this sighting may be associated with this development.

Canada Goose *Branta canadensis* (Linnaeus)

Photo: Martin Steer

Recent introduction

This species appears to have been introduced to Clare Island in 2014, and four adults were first recorded there on 12 March 2014. These birds now appear to have joined a free-flying flock of both Canada and feral greylag geese that frequent the area around Roonah Lough on the nearby mainland.

Barnacle Goose *Branta leucopsis* (Bechstein)

Photo: R.T. Mills

Annual winter visitor in the past. Now probably a regular winter and spring visitor in small numbers (PW)

Ussher (1912) said that this species was a 'common winter visitor in flocks' in 'March and April and again in September, October, and November'. Ruttledge (1950) quotes P. McMahon that '70 to 100 winter'.

It appears that between 11 and 100 were found on Clare Island during the winter atlas (November 1981—February 1984) (Lack 1986). However, barnacle geese were not found on Clare Island during the survey in the spring of 1988 (Walsh and Merne 1988), nor in any of the regular subsequent surveys. Ruttledge (1994) does not include Clare Island in his review of the distribution of this goose in Galway and Mayo, nor was it detected in the most recent atlas survey (Balmer *et al.* 2013). Therefore, it appears that it may have departed from Clare Island as a regular visitor sometime between 1984 and 1988.

However, a flock of 13 was found on Kinatevdilla on 27 January 2006 and a flock of 40 was also found on the island by Malachy Mangan on 16 March 2018 (www.Irishbirding.com). A flock of about 25 birds was seen at Tore More in the spring of 2018 (PW). Patsy Winters believes that it is a regular winter visitor in small numbers, favouring one field in Tore More where it could easily be overlooked.

The absence of regular grazing geese from Clare Island is interesting as obligate herbivorous bird species generally, in terms of biomass, form a very small percentage of the estimated total of breeding land birds on the island (see Chapter 4). The very large increase in sheep numbers may be, at least, a partial explanation, though equally, in the case of geese, grazing could enable more efficient foraging by these birds.

Common Shelduck *Tadorna tadorna* (Linnaeus)

One record

Not included in Ussher's (1912) inventory for Clare Island, nor in any subsequent survey until one pair was seen in suitable breeding habitat on 27 May 2010 (Balmer *et al.* 2013). While this may indicate probable breeding, there were no subsequent sightings of any kind since that observation, suggesting that a nest may not have been present (see Ferguson-Lees *et al.* 2011).

Eurasian Wigeon *Anas penelope* Linnaeus

No recent record

Though omitting any mention of this duck in his 'comparative table', Ussher (1912) describes this species as a 'common winter visitor to Clare Island'. However, there appears to have been no subsequent sightings (e.g. Lack 1986; Balmer *et al.* 2013).

Eurasian Teal *Anas crecca* Linnaeus

Only one recent record

'Common on Clare Island (McCabe), but not observed there in summer' (Ussher 1912). We have not been able to trace any subsequent winter records and neither of the winter atlas surveys detected teal on Clare Island (Lack 1986; Balmer *et al.* 2013).

A possible breeding pair was recorded by Jackie Hunt in April 2008 (per EM).

Mallard *Anas platyrhynchos* Linnaeus

Resident; breeds in small numbers

Mallard were found to be breeding at Craigmore (Ussher 1912), and subsequent surveys found the species to be resident though in relatively small numbers (see Chapter 5). This duck has also been recorded in the winter (Lack 1986; Balmer *et al.* 2013).[1]

Red-breasted Merganser *Mergus serrator* Linnaeus

No recent record

Ussher (1912) states that 'in winter flocks appear' in Clew Bay and 'visit' Clare Island. However, this observation is not included in the summary table (Ussher 1912). There appears to have been no further sightings (e.g. Ruttledge 1950, 1994;

[1] A drake pintail *Anas acuta* Linnaeus was seen by Eoin McGreal at a lochán in Strake on 3 May 2019, the first island record.

Sharrock 1976; Lack 1986; Balmer *et al.* 2013); (see Chapter 5).

Family Tetraonidae

Red Grouse *Lagopus lagopus* (Linnaeus)

Now extinct on Clare Island

This species was not found during the first survey 1909–11 (Ussher 1912), but there was a confident assertion that the red grouse was once plentiful on the island and was hunted, particularly in winter, by the resident golden eagle (Ussher 1912; see also Ruttledge 1994).

The demise of the red grouse probably coincided with the 'degradation of the heather communities', which is well-described by Doyle and Foss (1986). Ryle (2013) states that 'the most significant vegetation changes that have occurred on Clare Island have been to the *Calluna* formation'. This would have been the main habitat of red grouse on Clare Island, and Doyle and Foss (1986) quote an 83-year-old farmer who stated that 'sheep farming was made difficult in the early part of the century because animals could not be seen in the tall heather sward on the mountains'. This habitat, which Praeger (1911) stated extended right up to the summit of Croaghmore, has been eliminated (Doyle and Foss 1986; Ryle 2013). However, *Calluna* heather has been shown to provide 'food, shelter and cover' for red grouse; and this plant is believed to form 'at least 90% of their diet for most of the year' (Savory 1978). Therefore, with the degradation and almost complete elimination of the *Calluna* formation, the extinction of the red grouse on Clare Island was inevitable.

The taxonomic status of the red grouse in Ireland remains unresolved (e.g. Ruttledge 1975; Hutchinson 1989; Freeland *et al.* 2007; McMahon *et al.* 2012; Cummins *et al.* 2010, 2015; Murray *et al.* 2013; Mayer-Lucht *et al.* 2016). The red grouse would surely have been among the first waves of post-glacial recolonisation of Ireland by birds (Kelly 2008; Kelly *et al.* 2014), and given the fact that Ireland has been an island and separate from Great Britain for 15,000 years it is perhaps surprising that the genetic 'footprint' is not more discrete (but see Mayer-Lucht *et al.* 2016). Of course, as pointed out by many authors, 'birds and eggs of the British race, *Lagopus l. scoticus*, have, from time to time, been introduced' (Ruttledge 1975) and this may have obscured the expected differentiation of the respective races, though 'no apparent change has taken place in the characteristic colouring of the Irish birds' (Ruttledge 1975).

Family Phasianidae

Pheasant *Phasanius colchicus* Linnaeus

Recent introduction

The pheasant was not listed in Ussher's (1912) inventory or in subsequent surveys until that of Kelly which took place from 2–5 September 2005 (Kelly unpublished data). However, the pheasant was introduced to Clare Island in the 1960s, but it did not become established (PW). The species was also recorded during the atlas of 2007–11 (Balmer *et al.* 2013) and in the land bird census of 2013 (see Chapter 4). The pheasant was not detected during two surveys in 2003 (conducted from 13 to 15 July and 30 August) (Kelly unpublished data), so it appears that it was reintroduced to Clare Island in either 2004 or 2005.

Family Gaviidae

Red-throated Diver *Gavia stellata* (Pontoppidan)

Probably a regular winter visitor

This species was a 'late autumn and winter visitor to Clare Island' though was 'not so well known as the great northern diver' (Ussher 1912). Two were seen off the east end of Clare Island during the winter atlas (1981–84) and the species was also recorded during the atlas of 2007–11 (Lack 1986; Balmer *et al.* 2013). The red-throated diver is probably a regular visitor to the coastal waters around Clare Island from late autumn through to March.

Black-throated Diver *Gavia arctica* (Linnaeus)

Scarce visitor but almost certainly under-recorded

Not recorded by Ussher (1912) or in any of the winter atlas studies (Lack 1986; Balmer *et al.* 2013). However, two were recorded off Clare Island on 5 March 1984 (Ruttledge 1994).

Great Northern Diver *Gavia immer* (Brünnich)

Regular winter visitor to at least late April

A common late autumn and winter visitor from October to at least April to the waters around Clare Island (Ussher 1912; Kelly unpublished data).

Family Procellariidae
Fulmar *Fulmarus glacialis* (Linnaeus)

Resident since at least 1943 but now declining

The fulmar first colonised Ireland in 1911 (Fisher 1966). Although prospecting of the Clare Island cliffs appears to have commenced in 1915, breeding was not confirmed until 1943 (Fisher 1966). Lloyd (1984) argues convincingly that breeding probably commenced on the island's northern and western cliff ledges in 1935. The present status of the fulmar on Clare Island is summarised elsewhere (see Chapter 3).

Manx Shearwater *Puffinus puffinus* (Brünnich)

Regular in adjacent waters but breeding status remains uncertain

Although common in the waters around Clare Island (e.g. Ussher 1912; Lloyd 1984; Mackey *et al.* 2004; Kelly unpublished data) there is, to date, no confirmed record of breeding (see Chapter 3).

Family Hydrobatidae
European Storm Petrel *Hydrobates pelagicus* (Linnaeus)

Probably regular in adjacent waters but breeding status remains uncertain

The European storm petrel was reported to have been 'repeatedly caught at the Clare Island lighthouse in July and August, and probably breeds in escarpments west' (Ussher 1912). The lighthouse was decommissioned in 1965 and there have been no recent records of storm petrel on the island though it is possible that it breeds (Ussher 1912; Lloyd 1984; see also McGreal 2013). The storm petrel is common in the waters adjacent to Clare Island (see Mackey *et al.* 2004).

Leach's Storm Petrel *Oceanodroma leucorhoa* (Vieillot)

Probably regular in adjacent waters; probably breeds on the Bills Rocks

Though never recorded on Clare Island (Ussher 1912; Lloyd 1984; Kelly *et al.* unpublished data) it is likely to be at least a visitor to the surrounding waters, especially as it almost certainly breeds on the Bills Rocks located some 13km to the north-west (McGreal 2013). Probable breeding on the Bills Rocks was reconfirmed by David Cabot in 2018 (Cabot 2020). This species could, conceivably, also breed on Clare Island on the inaccessible western cliffs.

Much remains to be learned about the Leach's storm petrel in Ireland and a detailed review of its presence here, especially its movements around the coast in the spring and autumn, needs to be conducted. Sea watching from Clare Island under suitable weather conditions should yield sightings (Hutchinson 1989; Ruttledge 1994; Mackey *et al.* 2004; Parkin and Knox 2010; Jessopp and Kelly 2012; Balmer *et al.* 2013; Hobbs 2018 and references therein).

Family Sulidae
Gannet *Morus bassanus* (Linnaeus)

Colony established in 1978, sustained increase, and is common in adjacent waters

Although 'commonly to be seen fishing at sea round Clare Island' (Ussher 1912), it was not until 5 August 1978 that evidence of successful breeding was obtained by Michael Sweeney (Preston 1979; Lloyd 1984) 'when a juvenile was seen on a nest'. It is probable that a breeding pair was present from at least 1975 (Hutchinson 1989). Since then the colony has undergone remarkable growth to the most recently estimated total of 352 pairs (apparently occupied nests = AONs) (2017 estimate, McGreal unpublished data) (see Chapter 3).

Family Phalacrocoracidae
Great Cormorant *Phalacrocorax carbo* (Linnaeus)

First breeding record in the late 1980s, and is common in adjacent waters

Ussher (1912) noted that a few great cormorants had been seen 'flying past' Clare Island in spring, summer and autumn, 'but were not found breeding'. The cormorant was recorded as a breeding species on Clare Island in the 1988–91 atlas and Lloyd *et al.* (1991) indicated the size of the colony to be between 11 and 50 pairs. The cormorant colony on Clare Island is at Kinatevdilla and was estimated to hold 51 AON's over the 2015 to 2017 period (see Chapter 3).

Shag *Phalacrocorax aristotelis* (Linnaeus)

Resident but decreasing

Described by Ussher (1912) as 'resident and abundant'. The number of breeding pairs have undergone a recent decline since the 1999 estimate of 93 (AONs)—a decrease of >16% (see Chapter 3).

Family Ardeidae
Grey Heron *Ardea cinerea* Linnaeus

Regular visitor in small numbers; does not breed

Ussher (1912) believed that the grey heron only visited Clare Island as a 'straggler' after the breeding season. This species was not recorded by Barlee and Ruttledge (1945) but was seen on the island by Lloyd in June 1982 (Lloyd 1984). The grey heron was not recorded by Sharrock (1976), D'Arcy (1992), Gibbons *et al.* (1993), Forsyth and Allen (1997), Coombes (2003) or Winters (2006). One was seen on 29 May 2008 (Balmer *et al.* 2013). An adult was seen near Lough Avullin on 15 July 2000; another adult in the same place on 4 July 2003; a juvenile was seen near the Lough Avullin reed bed on 30 August 2003; one (unaged) flew in from the south at Portnakilly on 17 June 2013, but no sightings were obtained on 28–29 April 2000, 12–14 May 2001 or 2–5 September 2005 (Kelly unpublished data). One was seen at Portnakilly on 5 January 2016.

Family Accipitridae
White-tailed Eagle *Haliaeetus albicilla* (Linnaeus)

Photo: R.T. Mills

Very rare vagrant

One record: a female in second year plumage was shot on Clare Island on 27 November 1935 (Kennedy *et al.* 1954).

Hen Harrier *Circus cyaneus* (Linnaeus)

Vagrant

Two records: an adult male was seen by C.J. Patten 'about the middle of Clare Island' on 3 July 1910 (Ussher 1912) and a 'hen harrier' was seen on the island in May 1982 (Tony Whilde *in litt.* to Clare Lloyd); there seems to be no exact date, but it was sometime between 14–16 May 1982 (Lloyd 1984).

Sparrowhawk *Accipiter nisus* (Linnaeus)

Regular visitor; has bred

Described by Ussher (1912) as 'very uncommon' on Clare Island, the sparrowhawk was seen (single bird near the tower) in late April 1945 (Barlee and Ruttledge 1945) but it was not detected on the island during any of the three breeding atlas surveys (Sharrock 1976; Gibbons *et al.* 1993; Balmer *et al.* 2013). Ruttledge (1994) stated that this species does not breed on Clare Island but D'Arcy (1992) reported that this raptor bred on the island in 1989 in 'the wood at Ballytoohy'. In addition, Robert Cussen (1999, unpublished data) found that a pair of sparrowhawks successfully reared chicks every year between 1995 and 1999 in a nest in the alder *Alnus glutinosa* woodland in the Lassau valley south east of Ballytoohy above the coast at Portlea (see Ryle 2013). D'Arcy (1992) remarks that despite breeding on the island in 1989, the sparrowhawk was 'not seen at all in 1990' and there appears to have been no subsequent record of it nesting after Cussen's 1995–99 observations.

Incidental sightings include one during the winter atlas (Lack 1986), and although none was seen in late April 2000, early May 2001, August 2002, mid-July 2003 or mid-June 2013, a female was seen near the hotel on 30 August 2003, a male on 3 September 2005 (Football Pitch) and a female on the same date near the coast at Capnagower (Coombes unpublished data; Kelly unpublished data). Recoveries of sparrowhawks ringed in Ireland suggest that the species is effectively sedentary (Hutchinson 1989) so the autumn sightings listed above may have involved the dispersal of individuals that had nested on the nearby mainland (see Balmer *et al.* 2013) (see Chapters 4, 6 and 7).

Golden Eagle *Aquila chrysaetos* (Linnaeus)

Now extinct

Ussher (1912) gave a detailed account of the history of this species on Clare Island where the last individual was obtained on 14 October 1887 and sent to the National Museum in Dublin. T. Bourke Gaffney is quoted as stating that there was then (1878) 'an unbroken record of a pair occupying the same eyrie for the past century'. There are no records of this sedentary species (Parkin and Knox 2010) after 1887 (e.g. Ruttledge 1994).

Family Pandionidae

Osprey *Pandion haliaetus* (Linnaeus)

Rare vagrant

One record: one was seen on 15 June 2017 by Brian Bolger *et al.* (www.irishbirding.com).

Family Falconidae

Common Kestrel *Falco tinnunculus* Linnaeus

Resident; breeds

Ussher (1912) described the common kestrel as being 'common and resident on Clare Island' where it nested on many rocks (sea cliffs). Notably, three pairs of kestrels were estimated to be present during the late April, three-day survey of Barlee and Ruttledge (1945). However, Lloyd (1984) did not detect the kestrel in her 1982 survey and concluded that the species had become extinct on Clare Island—a view evidently supported by Ruttledge (1994) who stated that it 'no longer breeds on Clare Island'. Neither of these authors cite Sharrock (1976) where the kestrel was recorded as a breeding species on Clare Island. It was also 'seen' during the 1988–1991 breeding atlas (Gibbons *et al.* 1993) but was shown to be breeding by Balmer *et al.* (2013). Forsyth and Allen (1997) believed that at least one breeding pair was present during their surveys and a similar conclusion was reached by Kelly (unpublished data) in mid-July 2000 and mid-June 2013 (see Chapters 4 and 7). Breeding has been confirmed in recent years with two pairs recorded breeding in 2016 at coastal sites at Ballytoohy and Strake (EM), and possibly the same pair regularly seen at Tore More (PW). The latter nest site (i.e. at Strake) was also used in 2017 (EM). The kestrel was also recorded in winter by Lack (1986) and Balmer *et al.* (2013).

Merlin *Falco columbarius* Linnaeus

Regular visitor; does not breed

Ussher (1912) in his concluding 'Comparative Table of Birds of Clare Island, Achill, neighbouring islands, and adjoining mainland' places a question mark in the column relating to the possible breeding of the merlin on Clare Island. However, only one sighting, that of Patten in September 1910, is advanced in support of this possibility. This falcon was not found by Barlee and Ruttledge (1945); nor was it detected during the first breeding atlas survey (Sharrock 1976) or in subsequent atlas surveys (e.g. Gibbons *et al.* 1993; Balmer *et al.* 2013). Based on sightings of one male merlin on three occasions hunting over moorland near the lighthouse in late June 1982, Lloyd (1984) tentatively concluded that this species may be breeding on Clare Island. Subsequent records of Forsyth and Allen (1997), Cussen (unpublished data), Coombes (2003) and Kelly (unpublished data) show that the merlin is a scarce and irregular visitor to Clare Island. The merlin is well known to visit offshore islands; see especially Pierce and Roe (2006) and references therein, and Sharrock (1973). The merlin is a migratory species and ringing recoveries in Ireland show immigration from Iceland and—occasionally—Scotland (Hutchinson 1989; Ruttledge 1994; Wernham *et al.* 2002; Pierce and Roe 2006). It is probable therefore that many of the merlin sightings on Clare Island involve migrants, possibly from the Icelandic population (see Pierce and Roe 2006). The merlin is known to have bred on only one offshore Irish island (see Chapter 6).

Just like Professor Patten in 1910, Robert Cussen saw a male merlin take a rock pipit in Pigeon Cove in 'late summer 1998'. A male merlin was observed hunting meadow pipits at Lecarrow on 21 April 2016 (EM).

Gyr Falcon *Falco rusticolus* Linnaeus

Rare vagrant; two records

Ussher (1912) reported that on 10 March 1905 a light-keeper on Clare Island saw 'a white hawk the size of a sea-gull eating a curlew'; this appears to be an irrefutable sighting of a gyr falcon though Ussher (1912) only accorded it a question mark status in his summary table. It is relevant that Ruttledge (1994) notes that this falcon has been found in Mayo in the 'months of January to April, most often in March'.

A white phase, i.e. a Greenland type gyr falcon (BWP*i* V2; Mullarney *et al.* 2001), believed to be a female, was found dead and presented to Robert Cussen (unpublished data) in mid May 1994; it was photographed, and preserved by a taxidermist (Ruttledge 1994).

Peregrine Falcon *Falco peregrinus* Linnaeus

Resident

'Resident on Clare Island; one pair at least breed on the western cliffs' (Ussher 1912). This situation appears to have continued to the present day. The atlas surveys recorded it as breeding in 1968–72 (Sharrock 1976), 'seen' during 1988–1991 (Gibbons *et al.* 1993) and thought to be breeding in 2007–11 (Balmer *et al.* 2013). In addition, Cussen (unpublished data) found a nest in 1992 and 2001 with two chicks on each occasion but noted 'I suspect it bred on the island every year that I visited'. An observer needs to be quite intrepid to locate a nest on the very steep 800 feet high western cliffs of Clare Island. A breeding attempt was confirmed at this site during the 2017 national peregrine breeding survey, and though regularly seen on field trips to Clare Island, peregrine falcons are rarely observed away from the general area of the western cliffs (e.g. Coombes unpublished data; Kelly unpublished data).

Family Rallidae

Water Rail *Rallus aquaticus* Linnaeus

Resident but now extinct on Clare Island

The water rail was once a resident of Clare Island where it bred around Lough Avullin (Ussher 1912). However, none has been seen or heard since the first survey and it appears to have become extinct between 1911 and 1945 (see Barlee and Ruttledge 1945). A more detailed history is given in Chapter 5.

Corncrake *Crex crex* (Linnaeus)

Photo: R.T. Mills

Migrant; formerly bred but now apparently extinct on Clare Island; still a rare vagrant

Although a corncrake was seen in 2004, this migrant species is now apparently extinct as a breeding species on Clare Island (Winters 2006). It was 'very common' at one time (Ussher 1912), but the corncrake underwent a major decline in Mayo—particularly between 1988 and 1993—as detailed in Ruttledge's (1994 and references therein) comprehensive review. The corncrake was found to be breeding in the 1968 to 1972 atlas survey (Sharrock 1976) but was not noted by Lloyd (1984) who thought that it had been 'lost' since 1945; Sharrock (1976) was not cited. The corncrake was not found during the 1988–91 atlas survey (Gibbons *et al.* 1993) nor during that of 2007–11 (Balmer *et al.* 2013), although one was heard on Clare Island in June 2017 (PW).

Interestingly, it appears that the corncrake remained breeding on Inishbofin (*c.* 25 km south-west of Clare Island) for considerably longer, and in larger numbers, than it did on Clare Island, despite the population collapse in County Galway being much steeper than that in Mayo (Ruttledge 1994).

Moorhen *Gallinula chloropus* (Linnaeus)

Irregular breeder but difficult to survey

Has bred, but probably an irregular breeder of uncertain status due to the impenetrable reed bed at Lough Avullin. Moorhens were detected in the first survey: 'some are resident and breed on the small lakes in Clare Island' (Ussher 1912). One was heard at Maum Lough (*aka* Lough Avullin) by Barlee and Ruttledge (1945). One was seen during the first breeding atlas survey (Sharrock 1976), but the species was not found in any subsequent breeding or winter atlas surveys (Lack 1986; Gibbons *et al.* 1993; Balmer *et al.* 2013). Lloyd (1984) thought that it might have been overlooked in her late June 1982 survey.

Forsyth and Allen (1997) did not encounter the moorhen but D'Arcy (1992) reported a sighting in June 1990, although none in July 1989. An immature bird was seen in the marsh by the beach on 25 June 1999, and an adult, on the north side of the Lough Avullin reed bed at Maum, on 14 July 2003 (Kelly unpublished data) (see Chapter 5).

The main (harbour) beach looking south. This is the principal gathering area for gulls and for migrating wading birds e.g. ringed plover *Charadrius hiaticula* Linnaeus, turnstone *Arenaria interpres* (Linnaeus) and sanderling *Calidris alba* (Pallas) – the latter being very scarce. Pied wagtails *Motacilla alba* Linnaeus are to be seen on the beach, the adjacent areas of dunes, short grass and along the road margins. Photo: T. Kelly

Family Haematopodidae

Oystercatcher *Haematopus ostrelagus* Linnaeus

Resident; breeds in small numbers

Winters (2006) recorded 82 oystercatchers in her survey, some of which were probably early migrants (e.g. Hutchinson 1989; Wallace *et al.* 2001) (see Chapter 5).

Family Charadriidae

Ringed Plover *Charadrius hiaticula* Linnaeus

Resident; breeds in small numbers, especially in the Beetle Head area

Occasional flocks of >10 individuals probably involve migrants (Winters 2006; Kelly unpublished data). A small flock winters regularly on the harbour beach at Fawnglass (see Chapter 5).

Dotterel *Charadrius morinellus* Linnaeus

Rare vagrant

A single bird was seen by Eoin McGreal near the Signal Tower on 14 April 2011 (Fahy 2012).

European Golden Plover *Pluvialis apricaria* (Linnaeus)

Common and regular winter visitor in the past, but no recent records

According to Ussher (1912) 'flocks visit Clare Island in March and April, and again from August to December, most numerously in October and November'. Interestingly, there has been no subsequent record of a golden plover on the island, but the previous pattern of its presence is strongly suggestive of passage migration (e.g. Ruttledge 1994). The race involved is almost certainly the so-called northern golden plover *Pluvialis apricaria altifrons* (C.L. Brehm); most of the wintering golden plovers in Ireland breed in Iceland (Hutchinson 1989; Ruttledge 1994; Wernham *et al.* 2002). It appears that the migratory route through Clare Island that existed at the time of the first survey has been abandoned: there were no records of golden plovers in the winter atlas surveys of 1981–84 (Lack 1986) or 2007–11 (Balmer *et al.* 2013); this absence is perhaps the result of less hunting pressure on estuaries and the larger west of Ireland wetlands.

Northern Lapwing *Vanellus vanellus* (Linnaeus)

Migrant; small breeding population

Breeds in small numbers (Ussher 1912; Sharrock 1976; Gibbons *et al.* 1993; Balmer *et al.* 2013). However, following the very severe winter of 1944–45 it was 'neither seen nor heard' (Barlee and Ruttledge 1945). Since then it has been recorded during virtually every breeding season survey (e.g. Forsyth and Allen 1997; Winters 2006; Cussen unpublished data; Coombes unpublished data; Kelly unpublished data). In recent years breeding appears to be confined to two separate locations at Craigmore and Capnagower (see Chapter 5). Also seen in winter, for example, February and December 2018 (PW).

Family Scolopacidae
Sanderling *Calidris alba* (Pallas)

Apparent vagrant but probably more frequent than current records indicate

Not listed by Ussher (1912) as occurring on Clare Island, or detected there during the winter atlases (Lack 1986; Balmer *et al.* 2013). One was seen on the beach on 3 September 2005 (T.C. Kelly and R.E. Hutchinson).

Purple Sandpiper *Calidris maritima* (Brünnich)

Regular winter visitor; but few recent records

According to Ussher (1912) 'Barrington shot one on Clare Island' (31 March 1911) and 'McCabe says it was often seen there in winter'. Robert Cussen (unpublished data) saw it occasionally during intensive studies of the rocky intertidal zone, in the winter, in the mid-1990s.

Dunlin *Calidris alpina* (Linnaeus)

No recent records

Although listed—albeit ambiguously—in the comparative concluding table as a 'non-breeding' visitor to Clare Island (Ussher 1912), there appears to be no definite record of it ever having occurred there (Lack 1986; Balmer *et al.* 2013) (see Chapter 5).

Jack Snipe *Lymnocryptes minimus* (Brünnich)

No recent records

According to Ussher (1912) the jack snipe was a winter visitor to Clare Island. It appears that there have been no subsequent sightings. It must be borne in mind that apart from the atlas surveys, Clare Island appears to be very rarely visited by field ornithologists during the winter months and this small and elusive species could be easily overlooked. It was not found during either of the winter atlas surveys (Lack 1986; Balmer *et al.* 2013).

Common Snipe *Gallinago gallinago* (Linnaeus)

Has bred

Resident species; breeds in small numbers and overwinters (Lack 1986). This snipe has not been recorded in the most recent breeding atlas (Balmer *et al.* 2013) (see Chapter 5). This species is regularly encountered during the winter months (PW).

Woodcock *Scolopax rusticola* Linnaeus

Once a common and numerus winter visitor but there have been no records since 1911

'Visits Clare Island and Achill most numerously from the end of October to the end of December, though some remain until March' (Ussher 1912). Since then there has been no recorded sighting of a woodcock on Clare Island (Sharrock 1976; Lack 1986; Balmer *et al.* 2013). Large numbers of woodcock were shot in Co. Mayo at 'the famous Ashford covert at Cong – 11,265 over the 33-year period from 1905 to 1938' (Ruttledge 1994). It would be interesting to know if this level of mortality influenced its distribution elsewhere in Co. Mayo.

Eurasian Whimbrel *Numenius phaeopus* (Linnaeus)

Photo: R.T. Mills

Lough Creggan – west of Knockaveen. Clare Island does not have a large body of open water. Lough Creggan is too small to retain breeding ducks and so its main ornithological interest most likely lies (as with many of the other lochåns around the island) as an attractant of vagrant North American shorebirds such as the pectoral sandpiper *Calidris melanotos* (Vieillot) in the autumn. Photo: T. Kelly

Passage migrant in spring and autumn

This species is a typical spring and autumn passage migrant with perhaps more birds in the spring than during the return migration in the autumn (Ussher 1912; Hutchinson 1989; Ruttledge 1994; see also Alves *et al.* 2016). In the current survey birds were seen in late April 2000, May 2001, late August 2002 and early September 2005; whimbrels were not seen during the mid-June survey of 2013 (Coombes unpublished data; Kelly unpublished data). The whimbrel was not recorded during any of the atlas surveys (Sharrock 1976; Lack 1986; Gibbons *et al.* 1993; Balmer *et al.* 2013).

In recent surveys the maximum number seen was 15 on 29 April 2000 (Coombes unpublished data). However, McGeehan (2018) shows a very large flock (by known Clare Island standards) of ≥ 21 birds which he photographed on 18 May 2014. It appears that during very heavy rain, and poor visibility, large numbers of whimbrels interrupted their journey northwards and 'sat out' the difficult conditions until the storm passed (McGeehan 2018).

On 3 September 2005 three whimbrels (two adults and a juvenile) were seen foraging in the littoral at Kinincorra, effectively a very small storm beach with perfectly rounded 'boulders' (9–22cm in diameter) closely grouped together like billiard balls. The adult whimbrels, but not the juvenile, were systematically and elegantly searching for prey under the cobbles, their decurved bill being perfectly adapted to exploiting this habitat; each whimbrel would slide its bill down the curved side of the cobble and probe the soft substrate underneath. A small crab was one identified prey item thus captured.

Eurasian Curlew *Numenius arquata* (Linnaeus)

Common passage migrant; does not breed

The curlew is a common visitor to Clare Island but mostly in small numbers, though a flock of 50 was recorded on 18 July (year not given, Ussher 1912). Between 10 and 99 were recorded by Lack (1986). This species was not reported by Barlee and Ruttledge (1945) but was recorded in both May and June 1982 (Lloyd 1984). It was apparently not seen in the June 1989 and July 1990 surveys of D'Arcy (1992) and it was not recorded by Forsyth and Allen (1997). Coombes (unpublished data) saw two curlews on 30 April 2000; curlews were seen in July 2000, August 2002, July 2003 and September 2005. Patsy Winters recorded one individual in June 2006. Curlews were not seen during the mid-June survey in 2013 (Kelly unpublished data). A total of 61 was recorded during a Non-Estuarine Waterbird Survey (NEWS) along the southern shore on 5 January 2016. One was observed in Lecarrow in December 2018 (PW). The curlew does not breed on Clare Island (e.g. Ussher 1912; Sharrock 1976; Balmer *et al.* 2013).

A flock of ten curlews was seen feeding in an unmown field at Capnagower accompanied by a small flock of starlings; notably the latter were entirely uninhibited by the long grass (Kelly unpublished data).

Common Sandpiper *Acitis hypoleucos* (Linnaeus)

Photo: R.T. Mills

Migrant; breeds

This migratory species still breeds in very small numbers on the island's south coast, as it did during the first survey (Ussher 1912) (see Chapter 5).

Common Redshank *Tringa totanus* (Linnaeus)

Once a regular visitor; now very scarce

According to Ussher (1912) the redshank was 'met with on Clare Island in March (Barrington); in small parties along the shore in July; increased in September; and in numbers in the fields in December (Patten)'. A very different picture emerges since then; the redshank was not discovered on Clare Island during any of the breeding or winter atlas surveys (Sharrock 1976; Lack 1986; Gibbons *et al.* 1993; Balmer *et al.* 2013). It was not reported by Lloyd (1984), D'Arcy (1992), Forsyth and Allen (1997) or Winters (2006). One was present near the beach on 26 August 2002 (Kelly unpublished data).

There is no obvious explanation as to why there is now an almost total absence of redshank compared to that found in the first study (e.g. see Ruttledge 1950).

Turnstone *Arenaria interpres* (Linnaeus)

Scarce passage migrant

Ussher (1912) quotes Patten who 'saw a few on Clare Island, in March and July, and flocks in September'. However, this species was not detected in any of the atlas surveys (see references under common redshank). Therefore, the only recent record appears to be a group of five, with eight ringed plovers, on the beach on 3 September 2005 (Kelly unpublished data) (see also Wallace *et al.* 2001).

The main harbour and pier at Clare Island. Gulls including the rare glaucous gull *Larus hyperboreus* Gunnerus, as well as terns visit this location. Seabirds including manx shearwaters *Puffinus puffinus* (Brünnich), arctic *Stercorarius parasiticus* (Linnaeus) and pomarine skuas *Stercorcarius pomarinus* (Temminck), pass through the sound with the mainland – from north to south – in large numbers in stormy conditions with strong south westerly winds. Photo: T. Kelly

Family Stercorariidae

Pomarine Skua *Stercorcarius pomarinus* (Temminck)

Scarce passage migrant

Scarce passage migrant; unknown from Clare Island during the first survey (Ussher 1912). Two separate birds were seen from the point at Kinincorra on 4 September 2005 (Kelly unpublished data). One was seen on 29 June 2013 by Jimmy Macdonald (www.irishbirding.com).

Arctic Skua *Stercorarius parasiticus* (Linnaeus)

Probably a regular passage migrant

Passage migrant; unknown from Clare Island during the first survey (Ussher 1912). One dark-phase bird passed south on 3 September 2005 and seven individuals flew south on 4 September 2005.

Pomarine/Arctic Skua *Stercorarius pomarinus/parasiticus*

One, not specifically identified, seen from a cliff to the north of the lighthouse, was flying among other seabirds feeding in the sea to the west of Clare Island on 18 June 2013 (T.C. Kelly and R.E. Hutchinson) (see Chapter 3).

Long-tailed Skua *Stercorarius longicaudus* Vieillot

Photo: R.T. Mills

Rare vagrant

One record: An adult bird, in Barrington's collection, 'was caught alive on a bog in Clare Island' on 15 June 1906 (Ussher 1912) (see also Ruttledge 1950, 1994).

Great Skua *Stercorarius skua* (Brünnich)

Commenced breeding in 2016[2]

The recent establishment of the great skua as a breeding species on Clare Island is one of the major developments in the history of its avifauna (Eoin McGreal pers. comm.; Martin Deithrick pers. comm.; PW) (see Chapter 3).

Family Laridae

Kittiwake *Rissa tridactyla* (Linnaeus)

Resident, but decreasing

Resident breeder, known since the first survey (Ussher 1912). The population size has decreased recently (see Chapter 3).

Black-headed Gull *Chroicocephalus ridibundus* (Linnaeus)

Passage migrant in small numbers

Occurs in small numbers mainly after the breeding season (Ussher 1912). Seen in June (Lloyd 1984); one adult and one juvenile 14 July 2003; two adults off the beach on 3 September 2005 (Kelly unpublished data). Also occurs in winter (Lack 1986; Balmer *et al.* 2013).

Common Gull *Larus canus* Linnaeus

Breeds in small numbers

The common gull was first discovered to be breeding on Clare Island in 1945 (Barlee and Ruttledge 1945). Although the size of the breeding colony is much reduced since 1999 (Cussen *et al.* 1999) this gull still breeds on Clare Island (see Chapter 3).

---

[2] The great skua did not breed on Clare Island in 2019 (EM personal observations).

Lesser-Black backed Gull *Larus fuscus* Linnaeus

Larus fuscus graellsii A.E. Brehm

Migrant; has bred since 1969; does not occur in winter

The lesser black-backed gull was initially recorded breeding on Clare Island in 1969 during 'Operation Seafarer', the first seabird census of Britain and Ireland (Cramp *et al.* 1974). Numbers are currently increasing. This migratory gull does not appear to occur on Clare Island in winter (Lack 1986; Balmer *et al.* 2013) (see Chapter 3).

European Herring Gull *Larus argentatus* Pontoppidan

Larus agentatus argenteus C.L. Brehm

Resident and increasing

This gull is a resident on Clare Island and was present during the first survey (Ussher 1912). Its present status is summarised in Chapter 3. Approximately 30 herring gulls (and one adult lesser black-backed gull), which were seen feeding in fields on 4 September 2005, were pursuing ants; a very big ant-swarming event had been seen earlier that day (Kelly unpublished data).

Glaucous Gull *Larus hyperboreus* Gunnerus

Photo: R.T. Mills

Scarce occasional visitor

A minimum of three records: third-winter bird (*sensu* Mullarney *et al.* 2001) was seen with white-headed gulls on the beach on 13–14 July 2003 and—probably the same bird—coming into full adult plumage on 30 August 2003 at the same location. Adult bird on the beach on 2 September 2005 (Kelly unpublished data). One was seen on 18–19 December 2007; probably the same bird on 28 February 2008 and again on 9 April 2008 (Balmer *et al.* 2013; www.irishbirding.com).

Great Black-backed Gull *Larus marinus* Linnaeus

Resident and increasing

Ussher (1912) emphasised the large size of the colony on Clare Island and some adjacent seabird colonies found during the first survey. The great black-backed gull is a year-round resident on Clare Island and the number of breeding pairs appears to be increasing (see Chapter 3).

Large gatherings of great black-backed gulls were noted on the beach at Clare Island as follows: 104 on 28–29 April 2000; 116 on 14 July 2003; *ca.* 150 on 30 August 2003; *ca.* 150 on 2 September 2005. Only one adult was found there on 16 June 2013. The large gatherings (which also included herring gulls and hooded crows) appear to have been associated with the storage on the pier of bags of fish food for the salmon cages on the north side of the island. Daily movements were observed eastwards to the pier during the day, and westwards towards Beetle Head in the evening (PW).

Family Sternidae

Sandwich Tern *Sterna sandvicensis* Latham

Occasional, but probably a regular passage migrant

This species does not breed on Clare Island and was not noted during the first survey (Ussher 1912). It is a passage migrant in small numbers, mostly it would appear, in the months after the breeding season. Recorded in June 1982 (Lloyd 1984); 3 on 15 July 2000; 2 on 14 July 2003; 2 (1 adult and 1 juvenile) on 30 August 2003; 3 on 4 September 2005 (Kelly unpublished data).

Common Tern *Sterna hirundo* Linnaeus

Scarce visitor to adjacent waters

The only records are of a sighting in L78 during the first breeding atlas (1968–72) (Sharrock

1976), and another of a flying bird on 19 June 2010 (Balmer *et al.* 2013).

Arctic Tern *Sterna paradisaea* Pontoppidan

Once bred but is now extinct on Clare Island; passage migrant in low numbers

The small breeding colony on Kinatevdilla discovered during the first survey (Ussher 1912) had been abandoned by the time of Barlee and Ruttledge's (1945) survey. There is no subsequent record of this species breeding on Clare Island (see Chapter 3). Incidental records include one in late July 1990 (D'Arcy 1992); one off the beach on 14 July 2003, and one seen from the ferry, carrying food, on 15 July 2003 (Kelly unpublished data).

Common/Arctic Tern *Sterna hirundo/paradisaea*

A so-called 'comic tern', i.e. one that could not be confidently identified as either a common or arctic tern was seen off the beach on 13 July 2000 (Kelly unpublished data).

Family Alcidae
Common Guillemot *Uria aalge* (Pontoppidan)

Resident; perhaps decreasing

A breeding species with colonies mostly on the western cliffs (Ussher 1912; Cussen *et al.* 1999; Newton *et al.* 2015) (see Chapter 3).

Razorbill *Alca torda* Linnaeus

Resident; increasing

Breeding colonies occur on the western, south-western and northern cliffs (Ussher 1912; Cussen *et al.* 1999; Newton *et al.* 2015) (see Chapter 3).

Black Guillemot *Cepphus grylle* (Linnaeus)

Resident

A breeding species found mainly on the northern and southern cliffs (Ussher 1912; Cussen *et al.* 1999; Newton *et al.* 2015) (see Chapter 3).

Puffin *Fratercula arctica* (Linnaeus)

Migrant; breeds in small numbers[3]

The colony held ≤ 50 pairs in 1999 but the numbers now breeding on the island are unknown (Ussher 1912; Cussen *et al.* 1999) (EM pers. comm; see Chapter 3). The puffin is a migrant species and is absent in winter (Lack 1986; Balmer *et al.* 2013).

Family Columbidae
Rock Dove *Columba livia* J.F. Gmelin

Resident

According to Ussher (1912) 'numbers breed and are resident on the cliffs of Clare Island'. This is the situation today; two separate flocks—one of 22 birds and, later, one of 16 were seen on the island on 3 September 2005—but smaller flocks were seen during every field visit between 1989 and 2013 (Kelly unpublished data; Coombes unpublished data; Winters 2006) (see also Barlee and Ruttledge 1945; Sharrock 1976; Lloyd 1984; Lack 1986; Gibbons *et al.* 1993; Forsyth and Allen 1997; Balmer *et al.* 2013). Flocks of rock doves are often found in association with sheep feeding troughs (EM).

Although it is well known that the 'wild Common Pigeons' have continued to decline as feral birds increased, the former appeared to have been 'lost' from north and west Ireland (Parkin and Knox 2010), although this does not seem to be the case on Clare Island where 'near pure' forms constitute the majority, but not the only morph on the island (see also McGeehan 2014). The Clare Island rock doves possess an immaculate plumage and are of a slightly different size and shape to the 'town' or 'feral pigeon' and appear to be identical to those described by McGeehan (2014) (see Chapters 4, 6 and 7).

Woodpigeon *Columba palumbus* Linnaeus

Very scarce visitor; does not breed

[3] No puffins could be found in a three-hour survey on 14 June 2019, which was conducted at a site where this species was present in 1999. The puffin may now be extinct as a breeding species on Clare Island (EM personal observations).

This species—then known as the 'Ring Dove'—was not found on Clare Island during the first survey (Ussher 1912) and has remained a very infrequent visitor since (Lloyd 1984; Forsyth and Allen 1997; Winters 2006). It was recorded during the first winter atlas (Lack 1986) but not the second (Balmer *et al.* 2013). Lloyd's (1984) record in May 1982 and the single sighting of a woodpigeon on 14 July 2003, and one seen by Seán Lysaght on 10 June 2016, appear to be the only reports of the species on Clare Island (Kelly unpublished data).

Collared Dove *Streptopelia decaocta* (Frivaldszky)

Scarce visitor; does not breed

There are only three records: one was seen on either 8 May or 18–20 July 1996 (Forsyth and Allen 1997); one was seen on the beach road near the harbour on the morning of 19 June 2013 and a further three were seen south of the harbour (at a shingle strand) later the same day (19.00 hrs); one was seen feeding in a farmyard at Capnagower on 22 October 2018. There was no evidence of breeding (see Chapters 4, 6 and 7).

Family Cuculidae
Common Cuckoo *Cuculus canorus* Linnaeus

Migrant; once a common and annual breeding species but now a very irregular visitor to Clare Island

The cuckoo appears to have become temporarily extinct on Clare Island sometime after the 1988–91 breeding atlas survey (Gibbons *et al.* 1993). One or two pairs bred in 1945 though the survey was conducted in late April (Barlee and Ruttledge 1945); considered to be breeding in both L68 and L78 during the first breeding atlas (Sharrock 1976) and several were heard in the late June 1982 survey of Lloyd (1984). Similarly, it was found in both atlas squares in the 1988–91 survey (Gibbons *et al.* 1993). However, no cuckoos were detected in the detailed studies of Forsyth and Allen (1997), Winters (2006), Balmer *et al.* (2013), Coombes (unpublished data) and Kelly (unpublished data), though Robert Cussen heard it every year (1995 to 1999) during his fieldwork on the island. One was heard in May 2012 and again in the early summer of 2017 and 2018 (PW), but whether it bred on these occasions is not known.

It is difficult to confirm breeding in the cuckoo, especially on an island where survey visits are usually quite short (Ferguson-Lees *et al.* 2011; see also Davies 2015). The cuckoo call is made by the male and it may or may not indicate the presence of a female, but the latter are usually difficult to see as they make themselves inconspicuous while identifying the nests into which they will lay an egg (e.g. Davies 2015).

Family Strigidae
Short-eared Owl *Asio flammeus* (Pontoppidan)

No recent record

According to Ussher (1912) the short-eared owl was a 'winter visitor to Clare Island', but there appears to be no subsequent record (Lack 1986; Balmer *et al.* 2013; see also Ruttledge 1994).

Family Apodidae
Common Swift *Apus apus* (Linnaeus)

Vagrant

According to Ussher (1912) 'on Clare Island a few swifts were observed, and six together, in June and July', but none was observed during any of the breeding atlas surveys (Sharrock 1976; Gibbons *et al.* 1993; Balmer *et al.* 2013). The swift was not encountered in the surveys of Coombes or Kelly (unpublished data) or that of Winters (2006), and the only recent record appears to be that of Lloyd (1984) in late June 1982.

Family Upupidae
Hoopoe *Upupa epops* Linnaeus

Rare vagrant

One record; one was shot on Clare Island on 25 March 1893 (Ussher 1912).

Family Corvidae
Chough *Pyrrhocorax pyrrhocorax* (Linnaeus)

Resident; breeds

'Clare Island may happily be called a sanctuary of the chough, where it is resident and common' (Ussher 1912) and this status continues today.

A flock of 30 choughs was seen by Pastor Lindner and Ussher in June 1910 (Forsyth and Allen 1997) and a similarly sized group was seen feeding 'in one small field' in late April (Barlee and Ruttledge 1945). Lloyd (1984) found five breeding pairs and reports that Ian Bullock in a late June survey dedicated to quantifying the numbers and distribution of the chough, found 'no less' than 25 breeding pairs. However, Forsyth and Allen (1997) counted 59 individuals 'north of Loughanaphuca' on 19 July 1997. Forsyth and Allen (1997) surmised that the number of breeding pairs 'would be of the order of twelve'. A total of 16 pairs was recorded on Clare Island during the 2002/03 national survey (Gray *et al.* 2003) while Eoin McGreal recorded a minimum of 12 pairs in 2016. Some recent large flocks include 20 on 30 April 2000 (Coombes unpublished data) and 24 on 3 September 2005 (Kelly unpublished data) and 32 in the summer months of 2006 (Winters 2006) (see Chapters 4, 6 and 7).

Magpie *Pica pica* (Linnaeus)

Newly arrived species; first bred in 1982

The magpie was not seen by any observer during the first survey but nonetheless was accorded the status of a straggler (Ussher 1912). It was not seen by Barlee and Ruttledge (1945) and does not appear to have been recorded during the first breeding atlas survey (Sharrock 1976). Therefore, Lloyd (1984) appears to have been the first observer to have recorded the magpie breeding on Clare Island.

However, Ruttledge (1994) stated in the context of the 'ever increasing and breeding' of this species in Co. Mayo, that on Clare Island 'at some time subsequent to 1945 a pair or two were breeding, certainly in the late 1960's and breeding continues'; unfortunately, no 'authority' is cited in support of this statement. The magpie was documented in L78 in Sharrock (1976) but this appears to have been in the eastern, (i.e. mainland) part of the square.

Nevertheless, the magpie is now established on Clare Island and it has been recorded there in the winter as well as the breeding season (see Chapters 4, 6 and 7).

Jackdaw *Corvus monedula* Linnaeus

Occasional visitor in small numbers; does not breed

According to Ussher (1912) the jackdaw was said to visit Clare Island in winter, and one was 'seen on the castle by Patten, 30 December' (1910). The species was not seen by Barlee and Ruttledge (1945), nor was it seen in 1982 (Lloyd 1984). However, Ruttledge (1994) states that 'at least one pair bred on Clare Island in 1982', though no supporting information is provided. Sharrock (1976) and Gibbons *et al.* (1993) shows its presence in L78 but apparently not on Clare Island. Likewise, the jackdaw was not recorded on Clare Island during the breeding seasons of 2008 to 2011 (Balmer *et al.* 2013). A total of 15 birds was recorded during the winter atlas (Lack 1986; Forsyth and Allen 1997), but the species was not recorded during the winter survey of Balmer *et al.* (2013) (see Chapters 4, 6, 7 and 8).

Recent sightings include two feeding in a field near Maum (14–16 April 1997, Forsyth and Allen 1997); two near the playing field on 12 May 2001; and one near the beach on 14 July 2003 (Kelly unpublished data).

Rook *Corvus frugilegus* Linnaeus

Occasional visitor; does not breed

Ussher (1912) knew well that although the rook did not then breed on Clare Island it did visit regularly from the mainland, sometimes in flocks, and quotes Praeger who saw a flock of '100 together' in July. Although Ussher (1912) suggests that most occurrences of rooks were from the late June to March period, Barlee and Ruttledge (1945) found three feeding on 24 April 1945 and Forsyth and Allen (1997) saw 'a few' feeding in fields during 14–16 April 1997. A total of 19 rooks (four of which were juveniles) were seen feeding in a field on the south side of Knockhaveen on 3 September 2005, and a single juvenile was seen flying in over the beach from the north on 5 September 2005 (Kelly unpublished data). These visitors probably originate from nearby Louisburgh where the number of breeding birds now exceeds 130 pairs (see Chapters 4, 6, 7 and 8).

Hooded Crow *Corvus cornix* Linnaeus

Resident; breeds

The hooded crow was well established as a resident species at the time of the first survey (Ussher 1912) and this status continues to the present day (Barlee and Ruttledge 1945; Sharrock 1976; Lloyd 1984; Lack 1986; Gibbons *et al.* 1993; Forsyth and Allen 1997; Winters 2006; Balmer *et al.* 2013).

The numbers of hooded crows increased from 2000 onwards, associated, it appears, with the storage of fish food on the pier in the harbour. For example, a flock of 16 juveniles was seen on 25 August 2002 and 17 (all of which appear to have been immature birds) were seen on 26 August 2002; and 26 (unaged) hooded crows were seen on 3 September 2005. All flocks were seen within 1km of the harbour area whereas numbers elsewhere on the island rarely exceed three birds, though Coombes (unpublished data) counted 25 in his 29 April 2000 survey and Winters (2006) also documented 25 in the Ballytoohy area of the island. The harbour flocks coincided with exceptionally large gatherings of gulls—especially great black-backed gulls—but by June 2013 the storage of the fish food on the pier had ceased and the gull and hooded crow flocks had disappeared (Kelly unpublished data) (see Chapters 4, 6, 7 and 8).

Raven *Corvus corax* Linnaeus

Resident; breeds

The raven is still to be seen virtually everywhere on Clare Island as it was at the time of the first survey (Ussher 1912). During a survey of breeding ravens in southwest Mayo in 2007 a total of five breeding pairs was recorded on Clare Island (McGreal 2007) and these sites were still being utilised in more recent years (see Chapters 4, 6, 7 and 8).

Family Regulidae

Goldcrest *Regulus regulus* (Linnaeus)

Has recently bred

Winters (2006) proved that this species bred on Clare Island for the first time in June 2006, when three chicks were reared. There are autumn records from 17 October 2008 and 22 October 2018 (Eoin McGreal), and a spring record from 16 March 2015 (Eleanor Keane) (www.irishbirding.com). One goldcrest was observed in Tore More in the summer of 2015. None was found to be breeding in June 2013 but the transect method used in that survey could have overlooked a small breeding population of goldcrests (see Chapters 4, 6 and 7).

Family Paridae

Coal Tit *Periparus ater* (Linnaeus)

Irregular straggler

Recorded by Eoin McGreal and Dermot Breen on 17 October 2008, and single birds recorded by Eoin McGreal at Ballytoohy and Maum on 8 October 2012. These birds probably belong to the Continental race *Periparus ater ater* Linnaeus as the endemic Irish taxon *P.a. hibernicus* (W. Ingram) is mostly sedentary though apparently it has been recorded on Cape Clear (Hutchinson 1989; but see also Sharrock 1973).

Family Alaudidae

Skylark *Alauda arvensis* Linnaeus

Resident; breeds

Still relatively common and widespread as it was during the first survey (Ussher 1912), and subsequently (Barlee and Ruttledge 1945; Lloyd 1984). A census in June 2006 showed the skylark to be an order of magnitude less abundant than the meadow pipit (see Chapters 4 and 7; Winters 2006; see also Hutchinson 1989). Although the skylark is almost unknown in winter on Cape Clear Island (Sharrock 1973), it was detected during both winter atlas surveys on Clare Island (Lack 1986; Balmer *et al.* 2013).

The skylark is well known to be one of the few passerine species, along with, for example, the swallow, that shows obvious though complex diurnal migration patterns, especially in the autumn (see Hutchinson 1989 for a review).

These daytime movements were detected on Clare Island by the lighthouse keepers in the nineteenth century (Barrington 1900; Ussher 1912) but recent information is lacking.

Family Hirundinidae

Sand Martin *Riparia riparia* (Linnaeus)

Migrant; has bred, but is now a scarce passage migrant[4]

Regarding the sand martin 'there is no record from Clare Island' (Ussher 1912). However, although not recorded in the first breeding atlas (Sharrock 1976), a colony of 20 nests was documented in tetrad X (L6884) of L68 in the second; this location is on the south side of the island about 1km from the abbey (Gibbons *et al.* 1993; Forsyth and Allen 1997). Subsequent surveys have not located any breeding colony (Winters 2006; Balmer *et al.* 2013). This species was not detected by Barlee and Ruttledge (1945), Lloyd (1984), D'Arcy (1992) or Forsyth and Allen (1997).

A single sand martin was seen on 12 May 2001 and another flew in from the northwest on 15 July 2003 (Kelly unpublished data).

Barn Swallow *Hirundo rustica* Linnaeus

Migrant; breeds widely on the island

Ussher (1912) stated that the swallow was a 'summer visitor to Clare Island in small numbers'. It appears to have become more numerous since then; a recent census showed it to be the third most numerous land bird species on Clare Island (see Chapter 4). The 'more than one pair' found by Ussher (1912) in June 1910 'in the arch of a huge sea-cave open at both ends, south of Granuaile's Castle' does not appear to have been revisited (see Lloyd 1984).

[4] Three occupied burrows were found by Eoin McGreal on 3 May 2019 at the beach at Fawnglass (L715855) and were still 'active' on 14 June. This is the first breeding record of the sand martin since the 1968–72 breeding atlas survey (Sharrock 1976).

House Martin *Delichon urbicum* (Linnaeus)

Photo: R.T. Mills

Migrant; does not breed; very scarce occasional visitor[5]

Pastor Lindner saw an individual on Clare Island in June 1910 (Ussher 1912). There appears to have been only two subsequent sightings—both in 1982—one in May and one in June (Lloyd 1984).

Family Phylloscopidae

Yellow-browed Warbler *Phylloscopus inornatus* (Blyth)

Photo: M. O'Briain

Vagrant; only one record

[5] Two pairs of house martins bred in 2019 – the nests were found on 14 June at Fawnglass (L713854). This is the first breeding record of the house martin on Clare Island (EM pers. obs.).

One was seen at Gorteen by Gavin Ó Sé on 8 October 2016 (www.irishbirding.com).

Common Chiffchaff *Phylloscopus collybita* (Vieillot)

Vagrant; only two confirmed records

Recorded at Ballytoohy on 17 October 2008, and one was seen at Gorteen on 8 October 2016 (www.irishbirding.com). It is remarkable that there have only been two confirmed sightings of this widely distributed migrant species; it does not breed on Clare Island and these records show that at least some migratory chiffchaffs visit the island in the autumn. This species appears to be increasing in the coastal margin of Mayo and north Galway – it breeds on Inishbofin and Inishturk (Balmer *et al.* 2013). Barrington (1900) did not obtain any specimen of chiffchaff from a western lighthouse and the migration of this species was heavily concentrated along the east coast (Ruttledge 1994).

Willow Warbler *Phylloscopus trochilus* (Linnaeus)

Photo: R.T. Mills

Migrant; breeds

Although 'not observed on Clare Island' (Ussher 1912), this warbler is now a well-established breeding species, having been first discovered there by Ian Forsyth and Pamela Allen in 1996 when 'three breeding territories were identified and another five in 1997' (Forsyth and Allen 1997). Subsequent surveys showed breeding willow warblers in both L68 and L78 (Winters 2006) (see Chapter 4). However, it appears that no breeding pairs were observed in L68 during the most recent atlas survey (Balmer *et al.* 2013).

A willow warbler—a well-known single brooded species—Evans *et al.* (2009), was seen carrying food on Clare Island on 14 July 2000 (Kelly pers. obs.), which may indicate some pairs here are double-brooded. The willow warbler, which is six times more abundant than the chiffchaff in Ireland (Crowe *et al.* 2014), has recently undergone a marked increase in the west and north-west of the country (Crowe *et al.* 2010; Balmer *et al.* 2013).

Family Sylviidae

Common Whitethroat *Sylvia communis* Latham

Migrant; once bred but is now absent[6]

According to Ussher (1912), the whitethroat was 'not infrequent during summer in the brakes and hedges of holdings on Clare Island'. This suggests that while it had an established breeding population the numbers might not have been high. Barlee and Ruttledge (1945) did not find it and while Forsyth and Allen (1997) thought their survey might have been too early (25–27 April 1945), others such as Balmer *et al.* (2013) state that they arrive in mid to late April (see also Ferguson-Lees *et al.* 2011). Sharrock (1973) showed that only a trickle arrived on Cape Clear in mid to late April and similar results were obtained by Boddy (1992). Lloyd (1984), who did not detect the species in her 1982 survey, concluded that the whitethroat had become extinct sometime after 1911 and before 1945. However, the first breeding atlas (1968–72) showed the whitethroat to be present on Clare Island (Sharrock 1976) in L68. None was found in subsequent surveys (e.g. Gibbons *et al.* 1993; Forsyth and Allen 1997; Winters 2006; Balmer *et al.* 2013). Ruttledge (1994) remarked that though it breeds on Inishbofin, 'it is now absent from Clare Island'.

[6] One was found at Capnagower on 14 June 2019 (EM pers. obs.).

While it might be argued that the disappearance of the whitethroat from Clare Island was related to the population crash incurred by this species due to desertification of the Sahel (Winstanley *et al.* 1974), its continued presence on nearby islands (Hutchinson and Ruttledge 1978; Ruttledge 1994) suggests that other factors, unique to Clare Island, are involved.

Family Locustellidae
Common Grasshopper Warbler *Locustella naevia* (Boddaert)

Vagrant; breeding status remains uncertain

This species was not recorded on Clare Island, or anywhere else in Co. Mayo during the first survey; it is not listed anywhere in the inventory (Ussher 1912). It appears to be difficult to prove breeding in the grasshopper warbler. Migrants are obvious as their loud reeling song is unmistakeable and readily detectable but, once they are mated, they only sing at dusk and dawn (Ferguson-Lees *et al.* 2011; Sam Bayley pers. comm.). Notably, however, as pointed out by Ruttledge (1994), unmated birds can sing all summer.

Over the period 1968–72, the grasshopper warbler was recorded breeding on Clare Island (Sharrock 1976), but not during the subsequent atlas surveys (Gibbons *et al.* 1993; Balmer *et al.* 2013), despite 'marked concentration of gains ... in the extreme west and southwest of Ireland' (Balmer *et al.* 2013). However, D'Arcy (1992) heard a grasshopper warbler singing in the same marshy locality (Fawnglass) in both 1989 (14–18 June) and 1990 (20–24 July) 'suggesting breeding rather than migration' (see also Ruttledge 1994). Hopefully, incontrovertible proof of its breeding on Clare Island will be obtained in the future. Breeding occurred for the first time on Cape Clear in 2011 (Bayley and Caffrey 2018; Sam Bayley pers. comm.) where previously the species was effectively unknown in summer (Sharrock 1973). (see Chapters 4, 6 and 7).

Family Acrocephalidae
Sedge Warbler *Acrocephalus schoenobaenus* (Linnaeus)

Photo: R.T. Mills

Migrant; breeds, but only at two locations

'Repeatedly identified in marshy ground near Clare Island harbour in June, 1910, and July, 1911' (Ussher 1912), as it is today (e.g. Coombes unpublished data; Kelly unpublished data; D'Arcy 1992; Forsyth and Allen 1997; Winters 2006; Balmer *et al.* 2013). Interestingly, the sedge warbler was not reported by Barlee and Ruttledge (1945) or by Lloyd (1984), leading the latter to conclude that the species had become extinct in the 1911–45 period. However, it is at least possible that Barlee and Ruttledge's (1945) visit was too early to detect this migrant species and Ruttledge (1950) even expressed doubt about Ussher's conclusion that it was breeding on Clare Island in 1910 and 1911.

The diagnostic 'tchurr' alarm call (Ferguson-Lees *et al.* 2011) has been heard on many occasions including 13 July 2003, and an adult was seen carrying food on 15 July 2000; one was seen on 25 August 2002 (Kelly unpublished data) (see Chapters 4, 6 and 7; Balmer *et al.* 2013). Sedge warblers were recorded at two locations in 2016: on 28 May (James Kilroy) and 10 June (Dr Líam Lysaght) respectively.

Family Bombycillidae
Bohemian Waxwing *Bombycilla garrulus* (Linnaeus)

Rare vagrant

Two records
An individual was observed in a garden in Tore More in December 2005 and again in early spring 2013.

Family Certhiidae
Eurasian Treecreeper *Certhia familiaris* Linnaeus

Formerly unknown, first recorded in 2007

First recorded at Tore More in summer 2007; and has been seen regularly since then (PW).

Family Troglodytidae
Wren *Troglodytes troglodytes* (Linnaeus)

Resident and breeds widely over the island

This species was common and widespread in the time of Ussher's (1912) survey, and it remains widespread today (see Chapters 4, 6 and 7).

Family Sturnidae
Common Starling *Sturnus vulgaris* Linnaeus

Breeds and occurs also in winter

At the time of the first survey the starling was described as a 'winter visitor in flocks to Clare Island' (Ussher 1912). It is not known for certain when it began to breed; Barlee and Ruttledge (1945) did not record it in their list and Ruttledge (1950), quoting P. McMahon, says that it visited Clare Island from November to March, 'but does not breed'. The starling was reported to be breeding during the first atlas (Sharrock 1976) though it was not found to be present during May and June 1982 (Lloyd 1984). It was found to be breeding in both L68 and L78 in the second atlas survey (Gibbons *et al.* 1993) and it was seen in July 1996 and April 1997 (Forsyth and Allen 1997). Winters (2006), Balmer *et al.* (2013), Coombes (unpublished data) and Kelly (unpublished data) also recorded this single brooded species on Clare Island during the breeding season. It is now the fourth most abundant breeding land bird on the island (see Chapter 4).

Flocks of starlings continue to winter on Clare Island (Lack 1986; Balmer *et al.* 2013). Two starlings were seen flying towards the island from the ferry on the 5 September 2005 (Kelly unpublished data) when about half-way to Roonah Pier.

Family Cinclidae
Dipper *Cinclus cinclus* (Linnaeus)

Photo: R.T. Mills

Extinct as a breeding species; very rare occasional visitor

The subspecies *Cinclus cinclus hibernicus* Hartert is the taxon involved, making it the only endemic Irish subspecies to be found on Clare Island (Ussher 1912; Ruttledge 1950), and whose breeding biology is well described (Smiddy *et al.* 1995).

The dipper was found to be breeding on Clare Island in the 1909–11 survey, during which several specimens were obtained for comparison with museum skins (Ussher 1912). However, the dipper was not seen again until 14–18 June 1989 when one individual was found (D'Arcy 1992; Ruttledge 1994). The dipper was not detected in any of the atlas surveys or those of Barlee and Ruttledge (1945), Lloyd (1984) or Winters (2006), although one was seen by John Feehan in November 2017. One wonders whether the collecting of specimens during 1909–11 contributed to their extinction.

Family Muscicapidae, Sub-Family Turdinae
Ring Ouzel *Turdus torquatus* Linnaeus

Not a valid member of the avifauna of Clare Island

Ussher (1912) accords this species 'straggler' status, but it was not detected by the first survey's team, or during any subsequent expedition, but see the important note in Ruttledge (1950) regarding a questioned sighting on Clare Island.

Blackbird *Turdus merula* Linnaeus

Resident; breeds widely on Clare Island

Although well established as a breeding species during the first survey, Ussher (1912) stated that 'this bird is reported to be a settler on Clare Island within the present century'. It is now the fifth most numerous land bird on Clare Island (see Chapter 4). There is an arrival of birds in late autumn and winter (Ussher 1912), as occurs elsewhere in Ireland (Hutchinson 1989).

Fieldfare *Turdus pilaris* Linnaeus

Former regular winter visitor and passage migrant

Ussher (1912) provides detailed information from several sources including the 'schedules of Hammond, the principal light-keeper', to show that this species was once a 'winter visitor in large numbers', but there does not seem to have been any subsequent records (Lack 1986; Balmer *et al.* 2013). Four birds were seen in a field at Portlea in late autumn in the late 1990's (RC). It is likely however, that the fieldfare is a regular winter visitor to Clare Island especially in very cold weather.

Song Thrush *Turdus philomelos* C.L.Brehm

Resident; breeds, and is present in winter; became extinct in 1945 after a period of severely cold weather in January of that year

Ussher (1912) said of the song thrush that 'a few breed on Clare Island'. However, after the severe preceding winter, Barlee and Ruttledge (1945) could not find any on the island during their late April survey. This species is well-known to be susceptible to the effects of prolonged frosts and snow (Sharrock and Fogden 1965; Borton 1987; Hutchinson 1989). Equally, severe winter weather in Britain and the Continent can lead to a large influx of song thrushes (Lack 1986; Hutchinson 1989; McGeehan 2018). This probably explains Patten's remarkable observation on 29 December 1910 of 'song-thrushes abundant on the beach, heather, bogs and cultivated parts' (Ussher 1912). Large numbers of living and dead 'thrushes' were observed by Mr Líam Winters during the infamous period of snow in 1947 (per PW). The result of a recent census shows the song thrush to be much less abundant on Clare Island than the blackbird (see Chapter 4).

Redwing *Turdus iliacus* Linnaeus

Former winter visitor; probably a regular spring and autumn passage migrant

Ussher (1912) observed that 'like the fieldfare, this species, together with its congeners and starlings, flies thither in times of snow, and perishes in great numbers'. The redwing was, at the time of the first survey, a winter visitor to Clare Island, but although it appears now to be absent (Lack 1986) or scarce (Balmer *et al.* 2013), it probably is much more regular during the October to March period—especially in periods of severe weather—than current records suggest. Large numbers were seen on the island during a period of very cold weather in December 2009 when many dead birds were also observed (PW).

There are two races: *T. iliacus iliacus* Linnaeus of Fenno-Scandia and *T. iliacus coburni* Sharpe—the so-called Icelandic redwing (see Parkin and Knox 2010 for details) (Ruttledge 1950; Hutchinson 1989). Although generally agreed to be largely inseparable in the field, see Killian Mullarney's (2017, www.irishbirding.com, sighting IB94607) very informative photographs, and notes, on the distinctive characteristics of the two races.

Mistle Thrush *Turdus viscivorus* Linnaeus

Formerly a very rare straggler; no recent records

Ussher (1912) regarded the mistle thrush as a 'mere straggler to Clare Island, met with in May and October'. However, there appears to have been no recent records (Lack 1986; Balmer *et al.* 2013). The mistle thrush first bred in Ireland in 1807 (Smiddy 2014), and although it has spread westwards from Co. Louth (its first breeding location), it does not appear to have colonised many offshore islands (see Chapter 6).

Family Muscicapidae, Sub-Family Muscicapinae
European Robin *Erithacus rubecula* (Linnaeus)

Resident

Resident and 'plentiful' on Clare Island, including in winter (Ussher 1912). The robin remains one of the most numerous land birds on Clare Island (see Winters 2006 and Chapters 4 and 6).

European Stonechat *Saxicola torquatus* (Linnaeus)

Resident; common and widely distributed

The stonechat was described by Ussher (1912) as 'resident in small numbers'. However, the most recent census estimated it to be among the top ten most numerous birds on the island (see Chapter 4). It is notable therefore that the stonechat was not seen by Barlee and Ruttledge (1945), which they explained most probably resulted from a period of exceptionally severe weather in late January of that year. This species is well known to be very vulnerable to the effects of a severe winter but being multi-brooded (Cummins and O'Halloran 2003; Ferguson-Lees *et al.* 2011) it seems to exhibit population stability in that it can 'bounce back' to equilibrium following a crash (e.g. McCann 2000). The stonechat was recorded by Sharrock (1976), Lloyd (1984), Lack (1986), Forsyth and Allen (1997), Winters (2006) and Balmer *et al.* (2013).

Northern Wheatear *Oenanthe oenanthe* (Linnaeus)

Migrant; breeds in large numbers

The recent (2013) census showed the wheatear to be the most numerous land bird on Clare Island (see Chapter 4), thus confirming the views of previous surveyors (e.g. D'Arcy 1992). In the absence of comparable surveys of other islands in the west and northwest, it is difficult to assert that the number of wheatears on Clare Island is exceptional, but this location would appear to be a good site at which to research the demography and migration (i.e. with 'electronic' tags) of the species. One question relating to the wheatear; is this long-distance migrant capable of direct flights from Clare Island to Northwest Africa (e.g. Kelly *et al.* 2014; Alves *et al.* 2016)?

It would be remarkable if the Greenland wheatear *Oenanthe oenanthe leucorhoa* (J.F. Gmelin) does not pass through Clare Island, especially in the autumn (e.g. Hutchinson 1989; Wallace *et al.* 2001).

Pied Flycatcher *Ficedula hypoleuca* (Pallas)

Vagrant; one record

One was seen in a conifer grove at Maum on 4 September 2005 (T.C. Kelly and R.E. Hutchinson).

Family Prunellidae
Dunnock *Prunella modularis* (Linnaeus)

Resident and widely distributed

'Resident and frequent' at the time of the 1909–11 survey (Ussher 1912), the dunnock proved to be one of the most numerous land birds on Clare Island during the recent (2013) census (see Chapter 4). Remarkably, the dunnock was proven to be absent from Inishbofin and Inishshark in Hutchinson and Ruttledge's (1978) survey; see also Ruttledge 1957; but see Balmer *et al.* 2013).

Family Passeridae
House Sparrow *Passer domesticus* (Linnaeus)

Scarce but increasing resident

Although once a 'common resident' (Ussher 1912), Barlee and Ruttledge (1945) could find only six pairs in their survey. Lloyd (1984) saw a 'few birds' near the harbour in 1982. This species was apparently not detected during the first breeding atlas (Sharrock 1976) but was found in L68 in the second (Gibbons *et al.* 1993). Only one sighting was made in 1996 near the harbour (Forsyth and Allen 1997), and the species was not detected on the island during the first winter atlas (Lack 1986) but was observed in both summer and winter during the combined breeding and winter atlases of 2007–11 (Balmer *et al*. 2013). The house sparrow was found to be very scarce during the breeding season census of 2013 (see Chapter 4) and in the census by Winters (2006). In more recent years the house sparrow has increased in numbers, particularly in Capnagower (EM) and more widely across the island (PW). The increase may be part of the general recovery of the species across Ireland (Kelly unpublished data).

Stream at Ballytoohy Beg which may have been one of those inhabited by breeding dippers *Cinclus cinclus* (Linnaeus) at the time of the first survey. The grey wagtail *Motacilla cinerea* Tunstall has been recorded at this and similar sites. Photo: T. Kelly

Tree Sparrow *Passer montanus* (Linnaeus)

Not known in the first survey; now resident with a contracting population

The earliest report of the tree sparrow on Clare Island appears to be during the first atlas (Sharrock 1976). Though not recorded in May and June 1982 (Lloyd 1984), it was seen during the 1981–84 winter atlas (Lack 1986) and in every consecutive survey including Balmer *et al*. (2013). The most recent census in 2013 showed it to be quite widespread and abundant (see Chapter 4), and much more numerous than the house sparrow, but a survey in 2018 found *montanus* to be confined to Kill (EM), with *domesticus* now the dominant species.

Family Motacillidae

Grey Wagtail *Motacilla cinerea* Tunstall

Irregular breeder; also present in winter

The grey wagtail appears to be an irregular breeder on Clare Island (Ussher 1912; Balmer *et al*. 2013) and many surveys (e.g. Barlee and Ruttledge 1945; Sharrock 1976; Lloyd 1984; Gibbons *et al*. 1993; Forsyth and Allen 1997; Winters 2006) did not detect it during the breeding season. However, Coombes (unpublished data) and Kelly (unpublished data) saw grey wagtails at two separate locations in late April 2000; one was seen on 13 July 2003 and evidence of probable breeding was obtained on 15 July 2008 following an earlier sighting on 25 May of the same year (Balmer *et al*. 2013). The grey wagtail was not encountered during the census of land birds (see Chapter 4). The grey wagtail on Clare Island seems to be an example of the 'in-out-effect' described by Diamond and May (1977) (see Chapter 7).

Pied Wagtail *Motacilla alba* Linnaeus

Resident in small numbers

The pied wagtail has been recorded in all breeding season surveys since the first in 1909–11 (Ussher 1912; Balmer *et al*. 2013). The most recent census of land birds showed that while it remains a well-established breeding species, it is not present in large numbers (see Chapter 4).

The harbour beach looking north. Rock pipits *Anthus petrosus* (Montagu) nest in the low-lying cliffs and feed in the adjacent littoral. This is perhaps one of the most interesting bird species on Clare Island. The rock pipit is an understudied species in Ireland and much remains to be learned of its biology, ecology and population dynamics. Of particular interest is its capacity to withstand the impact of storms and high waves on its ecotone habitat—at the edge of the land and sea. A small colony of sand martins *Riparia riparia* (Linnaeus) was discovered here in the summer of 2019 by Eoin McGreal. Photo: T. Kelly

Meadow Pipit *Anthus pratensis* (Linnaeus)

Resident; numerous and widely distributed

'The most generally distributed bird on Clare Island (Patten), being found almost to the mountain's top, and is present at all seasons': this is the description by Ussher (1912) and the status of the meadow pipit does not appear to have changed greatly since the first survey (Sharrock 1976; Gibbons *et al.* 1993; Forsyth and Allen 1997; Winters 2006; Balmer *et al.* 2013). The meadow pipit was also present during winter surveys (Lack 1986; Balmer *et al.* 2013). Interestingly however, in Barlee and Ruttledge's (1945) survey which followed an exceptionally cold January, the meadow pipit was found to be 'scarce'. The most recent land bird census estimated the meadow pipit to be the second most numerous species on Clare Island (see Chapter 4).

Rock Pipit *Anthus petrosus* (Montagu)

Photo: R.T. Mills

Resident; small numbers

Ussher (1912) described the rock pipit as a 'common resident on the shores of Clare Island'. It appears that its status has not changed greatly since the time of the first survey (see Chapter 4). However, the shorelines of Ireland have taken an enormous pummelling from winter storms in recent years (see Chapter 3), and the impact of these events on the rock pipit which occupies this uniquely exposed littoral habitat, throughout the year, does not appear to have been quantified; a preliminary survey of some such areas on Cape Clear Island, Co. Cork, in May and June 2014 suggests that a decrease had occurred (Kelly unpublished data).

Family Fringillidae

Chaffinch *Fringilla coelebs* Linnaeus

Winter visitor; breeds occasionally

Ussher (1912) describes the chaffinch as 'a winter visitor to Clare Island from October to March'. Subsequent winter atlas surveys confirm that this status continues (Lack 1986; Balmer *et al.* 2013). Ruttledge (1994) discusses the different races that winter in Counties Galway and Mayo, emphasising that the species is also a passage migrant. Interestingly, Ussher (1912) also refers to 'the preponderance of those in female plumage which arrive in autumn' on Clare Island.

According to Forsyth and Allen (1997) there was one breeding record on Clare Island in 1970 but it is not clear if this was part of the first atlas survey (Sharrock 1976). This would be the first time the chaffinch had bred on Clare Island. Lloyd (1984) reported that at least one pair was present, apparently breeding, near the harbour in May and June 1982. The chaffinch was reported to be breeding in the 1988–91 atlas (Gibbons *et al.* 1993). The chaffinch was not discovered during the breeding season surveys of Coombes (2000, unpublished data) or Kelly (1989–2005, unpublished data) or that of Winters (2006) or Balmer *et al.* (2013).

However, chaffinches were found to be breeding in small numbers during the most recent land bird census in 2013 (see Chapter 4). This apparent irregular breeding pattern of the chaffinch on Clare Island suggests that it may also be an example of the 'in-out-effect' (Diamond and May 1977) (see Chapter 7). The chaffinch was recorded in three 1km squares on 10–11 June 2016 (Dr Líam Lysaght, pers. comm.).

Greenfinch *Carduelis chloris* (Linnaeus)

Winter visitor; recent decrease

'Winter visitor to Clare Island, whence Barrington got specimens in December and January' (Ussher 1912). The greenfinch was also found during the first winter atlas (Lack 1986) but, interestingly, not in the 2007–11 survey (Balmer *et al.* 2013). The greenfinch has declined in Ireland (and Britain) in recent years (Crowe *et al.* 2010, 2014) and this may be the explanation for its decrease as a wintering bird on Clare Island (see Balmer *et al.* 2013). Ruttledge (1994) states that the greenfinch breeds on Clare Island but we cannot find any verification of this in the literature.

Goldfinch *Carduelis carduelis* (Linnaeus)

Winter visitor in small numbers; very rare in summer

Described by Ussher (1912) as an 'occasional winter visitor in small numbers to Clare Island'; this is effectively identical to its present status (Lack 1986; Balmer *et al.* 2013). A flock of approximately 20 birds was seen in October 2018 (PW).

During recent breeding seasons the goldfinch was seen on 9 April 2007 (Balmer *et al.* 2013), and one was seen in Strake, near the village, on 19 June 2013, but there was no evidence of breeding (Kelly unpublished data). One was seen at Lecarrow on 11 June 2016 by Colin Heaslip, Tim Ryle, Eamonn Delaney and Heather Lally.

Linnet *Carduelis cannabina* (Linnaeus)

Resident; breeds widely but appears to be absent or very scarce in winter

The linnet is a common and widespread species on Clare Island, especially in June and July (e.g. Ussher 1912; Balmer *et al.* 2013). Small flocks (maximum nine birds) were seen in September; adults were seen copulating on 30 August 2003 and an adult was seen feeding chicks on 4 September 2005 (Kelly unpublished data). The linnet has quite a prolonged breeding season (e.g. Ferguson-Lees *et al.* 2011) (see Chapters, 4, 6 and 7).

During winter, the linnet was not found in either of the atlas surveys (Lack 1986; Balmer *et al.* 2013). The absence of the linnet in winter on Clare Island is interesting, although this is not discussed by Ruttledge (1950, 1994). However, Ussher (1912) suggests that it is probable that the 'frequent flocks of "linnets" reported in winter and spring from Clare Island [...] are largely composed of twites'. Barrington (1900) discusses the difficulty of separating these two species and suggests that the twite and the linnet are 'freely confounded' and, interestingly, argues that 'errors of identification are more numerous in the case of the linnet than any other species' (see Ussher 1912).

Twite *Carduelis flavirostris* (Linnaeus)

Once a resident species, now extinct on Clare Island

The twite, at the time of the first survey, was a 'common resident' on Clare Island and was present in the winter months (Ussher 1912). Twite were recorded by Barlee and Ruttledge (1945) and the species was 'seen' in the first breeding atlas survey (Sharrock 1976). Four breeding territories were identified in 1996 but none were documented in 1997 (Forsyth and Allen 1997). Twite were present during the first winter atlas survey (between 16 and 67 in number) in the 1981–84 period (Lack 1986). However, none were found in winter, or in the breeding season, in the comparable 2007–11 surveys (Balmer *et al.* 2013). The twite has undergone a major decline in Ireland and is now listed as a Red Data species with only 54 to 110 breeding pairs remaining in Ireland (see McLoughlin 2011 for a detailed review; Colhoun and Cummins 2013). Dedicated searches for twite on Clare Island between May 2005 and July 2008 proved unsuccessful (McLoughlin and Cotton 2008). Twite were not recorded in 2000 (Coombes unpublished data) or in April and July 2000, May 2001, August 2002, July 2003, or September 2005 (Kelly unpublished data). The twite was not encountered during the recent June 2013 land bird census (see Chapter 4). However, Winters (2006) recorded the twite at several locations around the island in June 2006 and suggested that the Rural Environment Protection Scheme may explain the increase in the numbers observed.

Lesser Redpoll *Carduelis cabaret* (P.L.S. Müller)

Recent arrival; breeds

D'Arcy (1992) was the first to record the lesser redpoll on Clare Island in mid-June 1989; this taxon was not encountered in the first survey (Ussher 1912) or by Barlee and Ruttledge (1945), Lloyd (1984), Forsyth and Allen (1997) or in any of the atlas surveys including the most recent (Balmer *et al.* 2013). Robert Cussen found it on the island in the summer of 1999, and although no nest was found, at least two pairs were believed to be breeding during the land bird census of June

2013 (see Chapters 4, 6 and 7). Winters (2006) saw five lesser redpolls, four of the sightings were in a wood on the northern part of the island.

Reed bed in Fawnglass in 2013 – sedge warbler *Acrocephalus schoenobaenus* (Linnaeus) and reed bunting *Emberiza schoeniclus* (Linnaeus) bred at this and nearby sites in 2013. Photo: T. Kelly

Family Emberizidae, Tribe Emberizini

Snow Bunting *Plectrophenax nivalis* (Linnaeus)

Scarce winter visitor

This species, designated as a winter visitor and passage migrant by Ruttledge (1994), was recorded several times during the first survey, in October, December, twice in March, once in April and on 10 September (Ussher 1912). There appear to be very few recent records (Ruttledge 1950, 1994; Lack 1986; Balmer *et al.* 2013) reflecting perhaps the low number of ornithologists visiting Clare Island in winter. A photograph of a bird seen on 18 October 2005 was downloaded onto the Birdwatch Mayo website and a male was seen by Eoin McGreal near the Signal Tower on 5 April 2007. A snow bunting was seen in November 2009 in Tore More (PW).

Lapland Bunting *Calcarius lapponicus* (Linnaeus)

Vagrant; one record

A single bird was seen by John Murphy on 21 September 2010 (www.irishbirding.com). This species is probably a regular visitor to Clare Island in the autumn and may over-winter (see Wallace *et al.* 2001).

Yellowhammer *Emberiza citrinella* Linnaeus

Photo: R.T. Mills

Once abundant; this species is now extinct on Clare Island

Ussher (1912) described the yellowhammer as 'one of our commonest resident birds' and was present 'all over Clare Island from March to September'. However, by April 1945 only three were found by Barlee and Ruttledge (1945). None has been seen since. The yellowhammer has also disappeared from other Irish islands including Cape Clear and Rathlin Island (Ruttledge 1966; Hutchinson 1989; Kelly unpublished data). These changes in the status of insular species—as well as 'northern and western Ireland over the last 40 years'—are probably due to the 'almost complete loss of arable farming' (Balmer *et al.* 2013).

Reed Bunting *Emberiza schoeniculus* (Linnaeus)

Photo: R.T. Mills

Resident in small numbers

The reed bunting, since the first survey, has been consistently found in small numbers mostly in the Fawnglass and Capnagower areas of Clare Island, the only exception being during the 1988–91 breeding atlas (Gibbons *et al.* 1993) (see Chapter 4).

Corn Bunting *Emberiza calandra* (Linnaeus)

Photo: R.T. Mills

Extinct on Clare Island

The corn bunting was present on Clare Island during the first survey (Ussher 1912) though perhaps was not 'abundant' as suggested by Barlee and Ruttledge (1945). The latter did not find one individual of this species during their late April survey in 1945. Therefore, the corn bunting became extinct between 1911 and some year prior to 1945. A decline also occurred in populations on some islands off the west coast of Scotland in the mid-1930s—a change in status unconnected, apparently, with modifications in agriculture (Donald *et al.* 1994 and references therein). This decline was well before the general population decrease and extinction of the corn bunting in Ireland (Hutchinson 1989; Taylor and O'Halloran 2002).

The 1930s was a period during which marked warming occurred in Ireland and more generally in northwest Europe (Kelly and Smiddy in prep), and it is tempting to think that the simultaneous demise of the corn bunting in Scotland and on Clare Island was an example of the Moran Effect (e.g. Hudson and Cattadorie 1999; Koenig 2002). However, it was not synchronised (except perhaps locally) in Scotland, and the species survived well on Inishbofin until at least 1976 (Hutchinson and Ruttledge 1978; Hutchinson 1989).

It appears that the corn bunting may not have been that numerous on Clare Island in the 1909–11 survey; e.g. contrast the accounts of the corn bunting and yellowhammer in Ussher (1912). At least one individual was shot using a sampling procedure, which was the norm at that time, but as with this and other species, such removal may, possibly, have precipitated a demographic accident.

Acknowledgements
This study would not have been possible without the warm welcome and patience (and informative discussion) shown us by the people of Clare Island. Equally, this paper could not have been written without the information from the various atlas databases given us by Brian Caffrey of BirdWatch Ireland. Joe Doolan and Martin Deithrick helped with the great skua breeding record. The contributors of records to Joe Doolan's invaluable www.irishbirding.com have added interesting sightings that have expanded the inventory of species making up this systematic list. We would also like to acknowledge the unpublished contribution of Ian Forsyth and Pamela Allen, not least for the thoroughness of their fieldwork, ingenious summary and the referencing of the databases they searched. We would also like to thank the late Mick Hartnett, the late Dr Kieran McCarthy, Ricky Whelan and Sam Bayley, all of whom generously assisted with the completion of this project. Finally, we would like to thank the referees for constructive critiques of the manuscript.

REFERENCES

Alstrom, P., Rheindt, F.E., Zhang, R., Zhao, M., Wang, J., Zhu, X., Yin, C.G., Hao, Y., Ohlson, J., Jia, C., Prawiradialaga, D.M., Ericson, P.G.P., Lei, F. and Olsson, U. 2018 Complete species - level phylogeny of the leaf warbler (Aves: Phylloscopidae) radiation. *Molecular Phylogenetics and Evolution* **126**, 141–52.

Alves, J.A., Dias, M.R., Mendez, V., Katrinaardottir, B. and Gunnarsson, T.G. 2016 Very rapid long distance sea crossing by a migratory bird. *Scientific Reports* **6**, 38154; doi: 10.1038/srep38154.

Balmer, D.E., Gillings, S., Caffrey, B.J., Swann, R.L., Downie, I.S. and Fuller, R.J. 2013 *Bird Atlas*

2007–2011: the breeding and wintering birds of Britain and Ireland. Thetford. BTO Books.

Barlee, J. and Ruttledge, R.F. 1945 Notes on the present status of birds on Clare Island. *Irish Naturalists' Journal* **8**, 311–3.

Barrington, R.M. 1900 *The migration of birds*. London and Dublin. Porter and Ponsonby.

Barrowclough, G.R., Cracraft, J., Klica, J. and Zink, R.M. 2016 How many kinds of birds are there and why does it matter? *PLoS ONE* **11**, e0166307; doi: 10.1371/journal.pone.0166307.

Bayley, S. and Caffrey, B. 2018 *Cape Clear Bird Observatory. Annual Report 2017; Report Number 26*. Kilcoole. Birdwatch Ireland.

Betts, M. 1992 *Birds of Skokholm*. Wales. Dyfed Wildlife Trust.

BWP *Birds of the Western Palaearctic Interactive*. BWPi Version 2.

Boddy, M. 1992 Timing of whitethroat *Sylvia communis* arrival, breeding and moult at a coastal site in Lincolnshire. *Ringing & Migration* **13**, 65–72.

Borton, D.G. 1987 The breeding passerines of Cape Clear Island 1986. *Cape Clear Bird Observatory Report* **19**, 50–61.

Cabot, D., Cabot, R. and Viney, M. 2020 The breeding seabirds and other birds on the Bills Rocks, County Mayo. *Irish Birds* **42**, 55–62.

Colhoun, K. and Cummins, S. 2013 Birds of Conservation Concern in Ireland. *Irish Birds* **9**, 523–44.

Coombes, D. 2003 Clare Island. Birds noted on the island during a 3-day visit 29/4 to 1/5/2000. Unpublished report.

Craig, R.J., Suh, A., Wang, M. and Ellegren, H. 2018 Natural selection beyond genes: identification and analyses of evolutionary conserved elements in the genome of the collared flycatcher (*Ficedula albicollis*). *Molecular Ecology* **27**, 476–92.

Cramp, S., Bourne, W.R.P. and Saunders, D. 1974 *The seabirds of Britain and Ireland*. London. Collins.

Crowe, O., Coombes, R.H., Lysaght, L., O'Brien, C., Choudhury, K.R., Walsh, A.J., Wilson, J.H. and O'Halloran, J. 2010 Population trends of widespread breeding birds in the Republic of Ireland 1998–2008. *Bird Study* **57**, 267–80.

Crowe, O., Musgrove, A. and O'Halloran, J. 2014 Generating population estimates for common and widespread breeding birds in Ireland. *Bird Study* **61**, 82–90.

Cummins, S., Bleasdale, A., Douglas, C., Newton, S., O'Halloran, J. and Wilson, H.J. 2010 *The status of the red grouse in Ireland and the effects of landuse, habitat and habitat quality on their distribution. Results from the National Red Grouse Survey 2007–2008*. Irish Wildlife Manuals No. 50. Dublin. National Parks and Wildlife Service.

Cummins, S., Bleasdale, A., Douglas, C., Newton, S., O'Halloran, J. and Wilson, H.J. 2015 Densities and population estimates of red grouse *Lagopus lagopus scotica* in Ireland based on the 2006–2008 national survey. *Irish Birds* **10**, 197–210.

Cummins, S. and O'Halloran, J. 2003 The breeding biology of the stonechat *Saxicola torquata* in southwest Ireland. *Irish Birds* **7**, 177–86.

Cussen, R., Kelly, T.C., Hartnett, M. and Walsh, P.M. 1999 *Counts of breeding seabirds on Clare Island, Co. Mayo, 1999*. Dublin. Report to Royal Irish Academy.

D'Arcy, G. 1992 Fauna observed on Clare Island, Co. Mayo, June 14–18 1990 *Irish Naturalists' Journal* **24**, 33–4.

Davies, N. 2015 *Cuckoo: cheating by nature*. London. Bloomsbury.

Dempsey, E. and O'Clery, M. 2007 *Finding birds in Ireland. The complete guide*. Dublin. Gill and Macmillan.

Diamond, J. and May, R.M. 1977 Species turnover rates on islands: dependence on census interval. *Science* **197**, 266–70.

Donald, P.F., Wilson, J.D. and Shepherd, M. 1994 The decline of the corn bunting. *British Birds* **87**, 106–32.

Doyle, G.J. and Foss, P.J. 1986 A resurvey of the Clare Island flora. *Irish Naturalists' Journal* **22**, 85–9.

Dymond, J.N. 1980 *The birds of Lundy*. Plymouth. Norwood, Roose and Co. Ltd.

Dymond, J.N. 1991 *The birds of Fair Isle*. Edinburgh. Ritchie.

Ellegren, H. 2013 The evolutionary genomics of birds. *Annual Review of Ecology, Evolution and Systematics* **44**, 239–59.

Evans, K.L., Leech, D.I., Crick, H.Q.P., Greenwood, J.J.D. and Gaston, K.J. 2009 Latitudinal and seasonal patterns in clutch size of some single brooded British birds. *Bird Study* **56**, 75–85.

Fahy, K. 2012 Irish Rare Bird Report 2011 *Irish Birds* **9**, 447–93.

Ferguson-Lees, J., Castell, R. and Leech, D. 2011 *A field guide to monitoring nests*. Thetford. British Trust for Ornithology.

Fisher, J. 1966 The fulmar population of Britain and Ireland, 1959. *Bird Study* **13**, 5–76.

Forsyth, I. and Allen, P. 1997 *Report to the Royal Irish Academy on the breeding birds of Clare Island, Co. Mayo, 1996 and 1997*. Dublin. Report to Royal Irish Academy.

Freeland, J.R., Andersson, S., Allen, D. and Looney, D. 2007 Museum samples provide novel insights into the taxonomy and genetic diversity of Irish red grouse. *Conservation Genetics* **8**, 695–703.

Gibbons, D.W., Reid, J.B. and Chapman. R.A. (eds) 1993 *The new atlas of breeding birds in Britain and Ireland: 1988–1991*. London. Poyser.

Gill, F.B. 2014 Species taxonomy of birds: which null hypothesis? *The Auk* **131**, 150–61.

Gill, F.B. and Donsker, D. (eds) 2018 IOC world bird list. Version 8.2. www.worldbirdnames.org.

Gray, N., Thomas, G., Trewby, M. and Newton, S.F. 2003 The status and distribution of choughs *Pyrrhocorax pyrrhocorax* in the Republic of Ireland 2002–2003. *Irish Birds* **7**, 147–57.

Hobbs, J. 2018 *A list of Irish birds*. Version.5.2, July 2018. www.southdublinbirds.com.

Hudson, P.J. and Cattadori, I.M. 1999 The Moran Effect: a cause of population synchrony. *Trends in Ecology and Evolution* **14**, 1–2.

Hutchinson, C.D. 1989 *Birds in Ireland*. Calton. Poyser.

Hutchinson, C.D. and Ruttledge, R.F. 1978 The birds of Inishbofin and Inishshark, Co. Galway. *Irish Birds* **1**, 211–24.

Jessopp, M. and Kelly, T.C. 2012 Open sea. In R. Nairn and J. O'Halloran (eds) *Bird habitats in Ireland*. 181–95. Cork. Collins Press.

Kapusta, A. and Suh, A. 2017 Evolution of bird genomes – a transposon's eye view. *Annals of the New York Academy of Sciences* **1389**, 164–8.

Kelly, T. 2008 The origin of the avifauna of Ireland. In J.L. Davenport, D.P. Sleeman and P.C. Woodman. (eds) *Mind the gap: post glacial colonisation of Ireland*. 97–107. Belfast. Irish Naturalists' Journal Special Supplement.

Kelly, T.C., Smiddy, P. and Graham, C.T. 2014 Origins of the Irish avifauna – a perspective from Island Biography Theory. In D.P. Sleeman, J. Carlsson and J.E.L. Carlsson (eds) *Mind the gap II*. new *insights into the Irish postglacial*. 44–52. Belfast. Irish Naturalists' Journal Special Supplement.

Kennedy, P.G., Ruttledge, R.F. and Scroope, C.F. 1954 *The birds of Ireland*. London and Edinburgh. Oliver and Boyd.

Koenig, W.D. 2002 Global patterns of environmental synchrony and the Moran Effect. *Ecography* **25**, 283–8.

Lack, P.C. 1986 *The atlas of wintering birds in Britain and Ireland*. Calton. Poyser.

Lloyd, C.S. 1984 The birds of Clare Island, Co. Mayo in June 1982. *Irish Naturalists' Journal* **21**, 212–6.

Lloyd, C.S., Tasker, M.L. and Partridge, K. 1991 *The status of seabirds in Britain and Ireland*. Calton. Poyser.

Lysaght, L. 2002 *An atlas of breeding birds of the Burren and the Aran Islands*. Dublin. BirdWatch Ireland.

Mackey, M., Ó Cadhla, O., Kelly, T.C., Aguilar de Soto, N. and Connolly, N. 2004 *Cetaceans and seabirds of Ireland's Atlantic margin*. Volume 1. Seabird distribution, density and abundance. Cork. Coastal and Marine Resource Centre.

Mayer-Lucht, Y., Mulder, K.P., James, M.C., McMahon, B.J., Buckley, K., Piertney, S.B. and Hoglund, J. 2016 Adaptive and neutral genetic differentiation among Scottish and endangered Irish red grouse (*Lagopus lagopus scotica*). *Conservation Genetics* **17**, 615–30.

Mayr, E. 1999 *Systematics and the origin of species from the viewpoint of a zoologist*. Harvard. Harvard University Press.

Mayr, G. 2011 Metaves, mirandornithes, strisores and other novelties – a critical review of higher-level phylogeny of neornithine birds. *Journal of Zoological Systematics and Evolutionary Research* **49**, 58–76.

McCann, K.S. 2000 The diversity-stability debate. *Nature* **405**, 228–33.

McGeehan, A. 2014 On the rocks – Ireland's wild rock doves. *Wings* (autumn) 2014, 6–7.

McGeehan, A. 2018 *To the ends of the earth: Ireland's place in bird migration*. Cork. Collins Press.

McGreal, E. 2007 The breeding status of the raven *Corvus corax* in southwest Mayo. *Irish Birds* **8**, 247–53.

McGreal, E. 2013 Probable breeding by Leach's storm petrel *Oceanodroma leucorrhoa* at Bills Rocks, County Mayo; results of a census of its breeding birds. *Irish Birds* **9**, 636–8.

McKay, B.D. and Zink, R.M. 2010 The causes of mitochondrial DNA tree paraphyly in birds. *Molecular Phylogenetics and Evolution* **54**, 647–50.

McLoughlin, D.T. 2011 *Management prescriptions for twite in Ireland*. Irish Wildlife Manuals No. 52. Dublin. National Parks and Wildlife Service.

McLoughlin, D.T. and Cotton, D. 2008 The status of the twite *Carduelis flavirostris* in Ireland 2008. *Irish Birds* **8**, 323–30.

McMahon, B.J., Johansson, M.P., Piertney, S.B., Buckley, K. and Hoglund, J. 2012 Genetic variation among endangered Irish red grouse (*Lagopus lagopus hibernicus*) populations: implications for conservation and management. *Conservation Genetics* **13**, 639–47.

Mullarney, K., Svensson, L., Zetterstrom, D. and Grant, P.J. 2001 *Collins bird guide*. London. Harper Collins.

Murray, T., Clotworthy, C. and Bleasdale, A. 2013 *A survey of red grouse* (Lagopus lagopus scoticus) *in the Owenduff/Nephin complex Special Protection Area, County Mayo*. Irish Wildlife Manual No. 77. Dublin. National Parks and Wildlife Service.

Newton, I. 2003 *The speciation and biogeography of birds*. London. Academic Press.

Newton, S.F., Lewis, L. and Trewby, M. 2015 *Results of a breeding survey of important cliff-nesting seabird colonies in Ireland 2015 – with an interim analysis of population changes*. Unpublished Report. Kilkcoole. Birdwatch Ireland.

Parkin, D.T. and Knox, A.T. 2010 *The status of birds in Britain and Ireland*. London. Christopher Helm.

Pierce, S. and Roe, J. 2006 Merlins *Falco columbarius* hunting at a small marine island. *Irish Birds* **8**, 79–84.

Praeger, R.L. 1911 Clare Island Survey, Part 10, Phanerogamia and Pteridophyta. *Proceedings of the Royal Irish Academy* **31**, 1–112.

Preston, K. 1979 Twenty sixth Irish bird report 1978. *Irish Birds* **1**, 413–99.

Prum, R.O., Berv, J.S., Dornberg, A., Field, D.S., Townsend, J.F., Lemmon, T.M. and Lemmon, A.R. 2015 A comprehensive phylogeny of birds (Aves) using targeted next generation DNA sequencing. *Nature*, **526**, 569–72.

Ruttledge, R.F. 1950 A list of the birds of the Counties of Galway and Mayo showing their status and distribution. *Proceedings of the Royal Irish Academy*. **52B**, 315–81.

Ruttledge, R.F. 1957 The birds of Inishbofin, Co. Galway. *Bird Study* **4**, 71–80.

Ruttledge, R.F. 1963 Migrant and other birds of Great Saltee, Co. Wexford. *Proceedings of the Royal Irish Academy* **63B**, 71–86.

Ruttledge, R.F. 1966 *Ireland's Birds*. London. Witherby.

Ruttledge, R.F. 1975 *A list of the birds of Ireland*. Dublin. National Museum of Ireland.

Ruttledge, R.F. 1994 *Birds in Counties Galway and Mayo*. Dublin. Irish Wildbird Conservancy.

Ruttledge, R.F. and Oglivie, M.A. 1979 The past and current status of the Greenland white-fronted goose in Ireland and Britain. *Irish Birds* **1**, 293–363.

Ryle, T. 2013 *Vegetation-environment interactions on Clare Island*. In D. Synnott (ed.) *New Survey of Clare Island*. Volume **7**, 27–178. Dublin. Royal Irish Academy.

Sangster, G. 2014 The application of species criteria in avian taxonomy and its implications for the debate of species concepts. *Biological Reviews* **89**, 199–214.

Sangster, G. 2018 *Integrative taxonomy of birds: the nature and delimitation of species*. In T. Tietze (ed) Bird species, how they arise, modify and vanish. 9–38. Switzerland. Springer.

Saunders, H. 1899 *Manual of British Birds*. London. Gurney and Jackson.

Savory, C.J. 1978 Food consumption of red grouse in relation to the age and productivity of heather. *Journal of Animal Ecology* **47**, 269–82.

Sharrock, J.T.R. 1973 *The natural history of Cape Clear Island*. Berkhamsted. Poyser.

Sharrock, J.T.R. 1976 *The atlas of breeding birds in Britain and Ireland*. Tring. British Trust for Ornithology.

Sharrock, J.T.R. and Fogden, S.C.L. 1965 The breeding passerines of Cape Clear 1965. *Cape Clear Bird Observatory Report* **7**, 42–49.

Smiddy, P. 2010 A systematic list of Irish Birds. *Irish Birds* **9**, 130–34.

Smiddy, P. 2014 Breeding birds in Ireland: success and failure among colonists. In D.P. Sleeman, J. Carlsson and J.E.L. Carlsson (eds) *Mind the gap II: new insights into the Irish postglacial*. 89–99. Belfast. Irish Naturalists' Journal Special Supplement.

Smiddy, P., O'Halloran, J., O'Mahony, B. and Taylor, A.J. 1995 The breeding biology of the dipper *Cinclus cinclus* in southwest Ireland *Bird Study* **42**, 76–81.

Taylor, A.J. and O'Halloran, J. 2002 The decline of the corn bunting *Milaria calandra* in the Republic of Ireland. *Biology and Environment: Proceedings of the Royal Irish Academy* **102B**, 165–75.

Tietze, T. (ed) 2018 *Bird species, how they arise, modify and vanish*. Switzerland. Springer.

Toews, D.P.L. 2015 Biological species and taxonomic species: will a new null hypothesis help? (A comment on Gill, 2014) *The Auk* **132**, 78–81.

Ussher, R.J. 1912 A biological survey of Clare Island in the County of Mayo, Ireland and of the adjoining district. *Proceedings of the Royal Irish Academy* **31**, Section 11, Part 20: 1–54.

Uva, V., Packert, M., Cibois, A., Fumagalli, L. and Roulin, A. 2018 Comparative molecular phylogeny of the barn owls and relatives (Family: Tytonidae) and their six major pleistocene radiations. *Molecular Phylogenetics and Evolution* **125**, 127–32.

Wallace, D.I.M., McGeehan, A. and Allen, D. 2001 Autumn migration in westernmost Donegal. *British Birds* **94**, 103–20.

Walsh, A. and Merne, O.J. 1988 Barnacle geese in Ireland, spring 1988. *Irish Birds* **5**, 539–50.

Wernham, C., Toms, M., Marchant, J., Clark, J., Siriwardina, G. and Baillie, S. (eds) 2002 *The migration atlas: movement of birds of Britain and Ireland*. London. Poyser.

Winstanley, D.R., Spencer, R.G. and Williamson, K. 1974 Where have all the whitethroats gone? *Bird Study* **24**, 1–14.

Winters, P. 2006 *A summer bird survey of Clare Island*. Unpublished M.Sc. thesis. Dublin. Centre for the Environment, Trinity College.

CHAPTER 3

THE SEABIRDS OF CLARE ISLAND

R.E. Cussen, E. McGreal, S. Warner and T.C. Kelly

ABSTRACT

The seabird fauna of Clare Island is described and compared with that recorded in the first survey. Currently, there are 14 species of seabird breeding on Clare Island compared to the eight noted in 1912 by Ussher and his co-workers. One species, the arctic tern *Sterna paradisaea* Pontoppidan, is no longer present, whereas the fulmar *Fulmarus glacialis* (Linnaeus), common gull *Larus canus* Linnaeus, lesser black-backed gull *Larus fuscus* Linnaeus, great cormorant *Phalacrocorax carbo* (Linnaeus), gannet *Morus bassanus* (Linnaeus) and, most recently (2015), the great skua *Stercorarius skua* (Brünnich)[1], have established breeding populations on the island at various times over the past century. In fact, remarkably, the species richness of seabirds on Clare Island has increased by 67 to 75% since 1912. The changing numerical abundances of each breeding seabird species are summarised and their distribution on the island's coastline is mapped from the results of both land-based and sea-based surveys conducted mostly in 1999 and 2015–17. Despite a marked recent decline of over 1,500 apparently occupied sites (AOS), the fulmar is the most numerous of the island's seabirds and its colony is the third largest of its kind in Ireland. Other species have also decreased, including the shag *Phalacrocorax aristotelis* (Linnaeus), kittiwake *Rissa tridactyla* (Linnaeus) and common gull, though probably for different reasons in each case. The gannet population has undergone a spectacular increase in recent times and now numbers some 352 pairs. These and related aspects of the seabird community breeding on Clare Island are discussed.

Introduction

As remarked by Wynne-Edwards (1964) the term 'seabird' is 'insusceptible of rigid definition'. This is because the marine bird group is drawn from four different orders in the Class Aves, of which three—the Procellariiformes, Pelecaniformes and Charadriiformes (Brooke 2002)—have representatives within the avifauna of Ireland (Hutchinson 1989). Interestingly, despite the enormous area covered by the oceans—some 70% of the globe—only 300 species or *c.* 3% of the world's total of bird species are seabirds. While these species are highly adapted, both structurally and physiologically, for the exploitation of this challenging habitat, they may be exceptionally vulnerable to predators when they come to land to breed (Goldstein 2002; Whittow 2002).

It is estimated that 96% of seabird species form colonies, usually on precipitously steep cliffs or

[1] The great skua did not breed on Clare Island in 2019 (Eoin McGreal – personal observation).

Cnoc Mór: the cliffs beneath the peak and stretching to the east and west hold most of the island's seabird colonies. The golden eagle *Aquila chrysaetos* (Linnaeus) had an eyrie here in the nineteenth century; today the peregrine falcon *Falco peregrinus* Linnaeus and ravens *Corvus corax* Linnaeus can be seen patrolling these cliffs. Photo: T. Kelly

remote offshore islands, which are mostly inaccessible to both mammalian and avian predators (Coulson 2002). Some seabird colonies are vast, holding in excess of one million birds (Croxall 1987), but few Irish colonies exceed 50,000 individuals (Mitchell *et al.* 2004; Newton 2012). Nevertheless, seabirds have variable, though exceptional, life histories characterised by very high annual adult survival but low reproductive output, a combination of demographic features that leads to relatively low resilience (Sibley *et al.* 2007; Dai *et al.* 2012; Neubauer *et al.* 2013) in their population dynamics; their capacity to recover or 'bounce back' to equilibrium following a population crash is relatively weak. In general, seabirds are at the 'slow' end of the 'fast–slow' continuum of avian life histories; seabirds are mostly classified as 'survivor species' (Saether and Bakke 2000). Population growth in most seabird species is driven more by adult survival than recruitment (Saether and Baake 2000; Clancy *et al.* 2006). Consequently, because seabirds are typically long-lived, they are especially vulnerable to additive adult, i.e. non-compensatory anthropogenic mortality (Burnham and Anderson 1984; Nichols 1991; Peron 2013) such as, for example, that which is caused by long-line fisheries in the Southern Oceans (e.g. Rivelan *et al.* 2010; Rolland *et al.* 2010; Phillips *et al.* 2016). The overall vulnerability of seabirds (Durant *et al.* 2009) to major disturbances in their environment leads to them being sometimes referred to as 'Canaries of the Sea'.

Gannets displaying. Photo: R.T. Mills

Great skua. Photo: R.T. Mills

Fulmar. Photo: R.T. Mills

The seabird fauna of Ireland is now well described, especially in terms of the numbers of breeding pairs occupying the numerous mainland and island colonies (Hutchinson 1989; Lloyd *et al.* 1991; Mitchell *et al.* 2004). In addition, detailed surveys are being conducted in the offshore zone where the exploitation of hydrocarbons, and increasingly, the location of wind farms, may pose threats in the future (e.g. Pollock *et al.* 1997; Mackey *et al.* 2004; Pollock and Barton 2008; Jessopp and Kelly 2012).

The numerical abundance of seabirds was not estimated during the first Clare Island survey, but a list of species was compiled (Ussher 1912). Subsequently, Barlee and Ruttledge (1945) and particularly Lloyd (1984) added considerably to the detail presented by Ussher (1912). However, the first comprehensive and optimally timed survey of the entire coastline of Clare Island and one that included a boat-based census of the cliffs below Cnoc Mór was conducted by Cussen *et al.* (1999). More recently the National Parks and Wildlife Service (Newton *et al.* 2015b; McGreal unpublished) are conducting a detailed census of the island's seabirds including a sea-based survey of the northern cliffs.

The aims of this study are: (i) to review the existing information on the numerical abundance and distribution of seabirds on Clare Island, (ii) to summarise the main changes in status of the species breeding on the island and (iii) to provide a preliminary thumbnail sketch of some features of the marine bird community on Clare Island in relation to biomass and feeding ecology.

Methods

Census of seabirds

The standard census units, optimally timed count dates, and methodologies recommended for Seabird 2000 (JNCC 1999; Walsh *et al.* 1995) were utilised in both the 1999 and 2015–17 surveys.

In 1999, *Larus* gulls were counted during 30–31 May (REC); land-based counts for all other species (fulmar *Fulmarus glacialis* (Linnaeus), kittiwake *Rissa tridactyla* (Linnaeus), common guillemot *Uria aalge* (Pontoppidan), razorbill *Alca torda* Linnaeus, black guillemot *Cepphus grylle* (Linnaeus), shag *Phalacrocorax aristotelis* (Linnaeus), great cormorant *Phalacrocorax carbo* (Linnaeus), puffin *Fratercula arctica* (Linnaeus) and gannet *Morus bassanus* (Linnaeus)) were made during 1–6 June, with the exception of the fulmar along parts of the western coastline and a segment beneath the Signal Tower, which were surveyed on 25 June (see Cussen *et al.* 1999 for details). The boat-based survey of 24 June 1999 included all areas not covered by the land-based counts.

In 2015 a boat-based survey was conducted on 10 June and supplementary land-based counts were carried out on 13 June. A complete black guillemot survey was carried out during 5 May 2017 using Seabird 2000 criteria (Mitchell *et al.* 2004); i.e. counts of adult birds before 09.00 hours (McGreal unpublished). The gannet colony has been censused annually since 2011. Census units (birds, nests, sites or territories; see Table 3) were generally counted singly (using binoculars or telescope). All available safe cliff-top vantage points were used, including distant views (maximum *c.* 1km) where this provided better angles for complete coverage. In addition, major subdivisions of the cliffs, compiled from sketches of individual cliff faces made in the field (see Cussen *et al.* 1999 for details) were used for the orderly completion of the census (i.e. ensuring that birds were not missed or double counted) and presentation of the count data (Fig. 1) (Cussen *et al.* 1999). These same subdivisions were adopted in 2015.

Considerable stretches of cliff on the north and west of the island are not visible from cliff-top vantage points and thus could not be surveyed in early June when the land-based counts in 1999 and 2015 took place. However, these were subsequently surveyed by boat. Due to weather and logistical difficulties the boat-based survey in 1999 was not attempted until 24 June when surveying was completed on the relevant sections. Weather conditions for most of the land-based

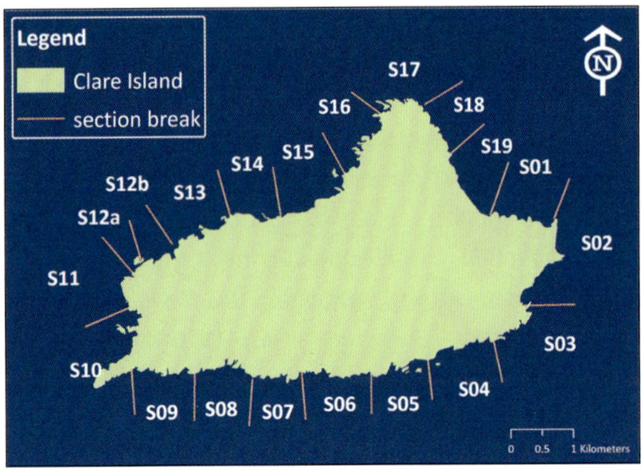

Fig. 1 Distribution of sections used to census seabirds on Clare Island in 1999 and 2015.

census and for the boat-based survey in both 1999 and 2015 were ideal, with light winds and clear skies. Further details of count times and weather for both the 1999 and 2015 counts are given on the Seabird 2000 data sheets (Cussen *et al.* 1999; Newton *et al.* 2015b).

Feeding ecology

Seabirds were classified as being either neritic (feeding primarily in the shallow inshore zone) or pelagic (feeding in the offshore zone in deep oceanic waters) (e.g. Mackey *et al.* 2004; Jessopp and Kelly 2012) (Table 1).

Biomass

The weights—both maximum and minimum—of seabird species breeding on Clare Island were obtained from the literature, especially Dunning (2008) and BWP*i* v2 (Table 2). The average weight of the male and female was added together and divided by two (note that if these data were not available the single average weight for those species, given in this literature, was used). The total biomass of each species was estimated by multiplying the average weight by the total number of individuals estimated to be present at the colony. It should be noted that for this calculation the total number of individual guillemots and razorbills counted in 1999 and 2015 was employed and the k conversion factor was not applied (Harris *et al.* 2015a).

Energy consumption

An estimate of the approximate energy consumption per day by the Clare Island seabird colony was based on a rate of 30% of the adult seabird biomass (see Furness 1978, 2007; Votier *et al.* 2004).

Results

Survey data

The results of the most recent surveys are summarised in Table 3, from which it can be seen that there have been substantial changes in the numerical abundance of the fulmar, shag, kittiwake and common gull *Larus canus* Linnaeus—all of which have decreased, while the gannet and the larger gulls (European herring gull *Larus argentatus* Pontoppidan, lesser black-backed gull *Larus fuscus* Linnaeus and great black-backed gull *Larus marinus* Linnaeus) have increased. The razorbill has also increased but the common guillemot colony appears to have declined since 1999, though this species is particularly difficult to census accurately, especially in the conditions that prevail on Clare Island where detailed and leisurely inspection of many of the ledges is not possible. The status of the puffin remains uncertain (see Figs. 2–8).

The 1999 survey showed the presence of 15,000+ individual seabirds, of which over half were fulmars (Table 3) whereas the 2015–17 census yielded a total of between 10,000 and 11,000 seabirds—a loss over the 16–17 year interval of in excess of 4,000 individual birds or approximately 30%. Moreover, the fulmar population was now less than half, indicating the severity of the decline in the numerical abundance of this species.

Biomass

The estimated biomass of the 1999 seabird population was approximately 12,040kg. On the basis that the 1999 population would consume some 30% of their body weight per day, the biomass of prey consumed per day would be of the order of 3,612kg. Although the population has decreased by 30% since 1999 (Table 3) the overall biomass has declined by 807kg or 7% as some of the largest species such as the gannet and great black-backed gull have increased over the interval. Based on the current estimate of approximately 11,000 seabirds on Clare Island, it appears that these may consume 3,321kg of prey per day.

Discussion

Changes in species composition

Fourteen species of seabird are now known to breed on Clare Island. This list represents 61% of the seabird fauna of Ireland, which is comprised of 23 species (Mitchell *et al.* 2004; Balmer *et al.* 2013). Although the arctic tern *Sterna paradisaea* Pontoppidan was present during the first survey (Ussher 1912), subsequent expeditions (e.g. Barlee

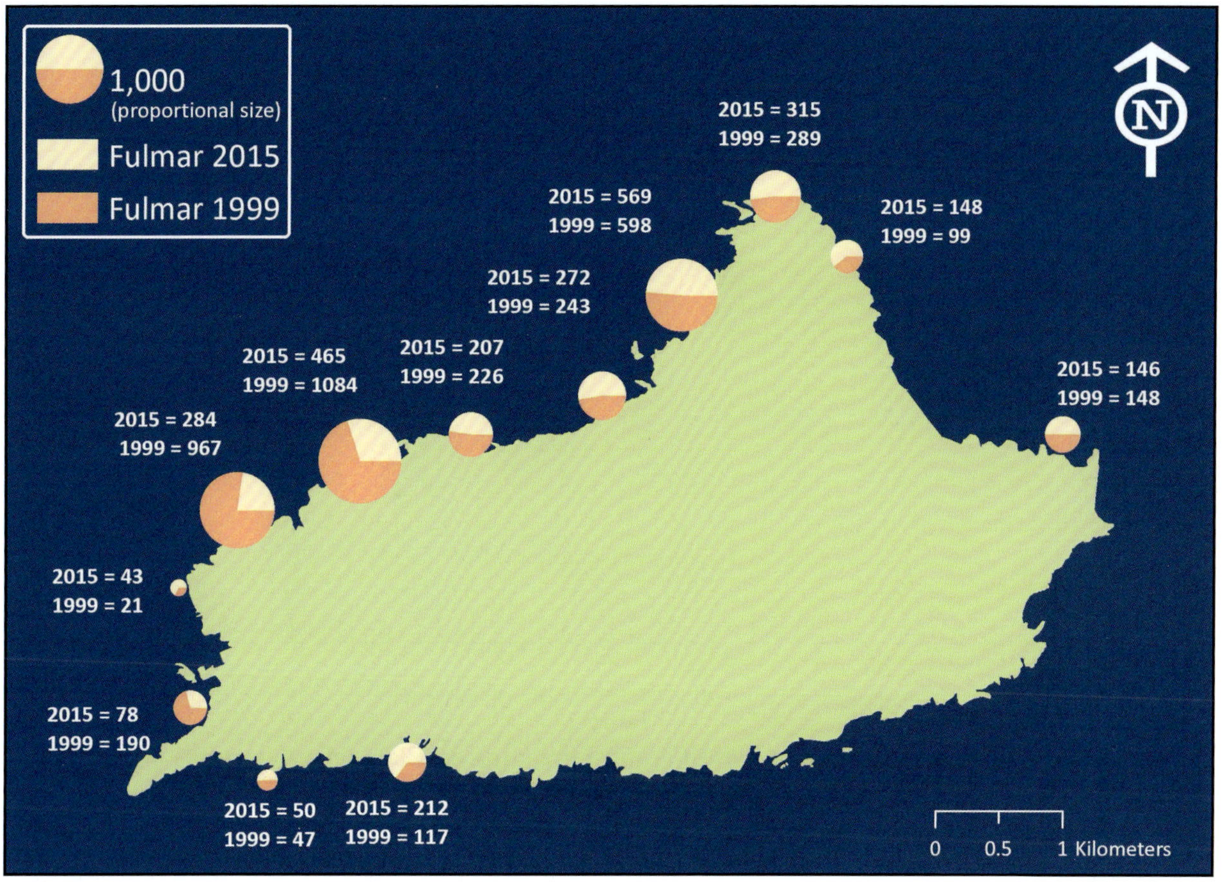

Fig. 2 Distribution and numerical abundance of the fulmar on Clare Island in 1999 and 2015.

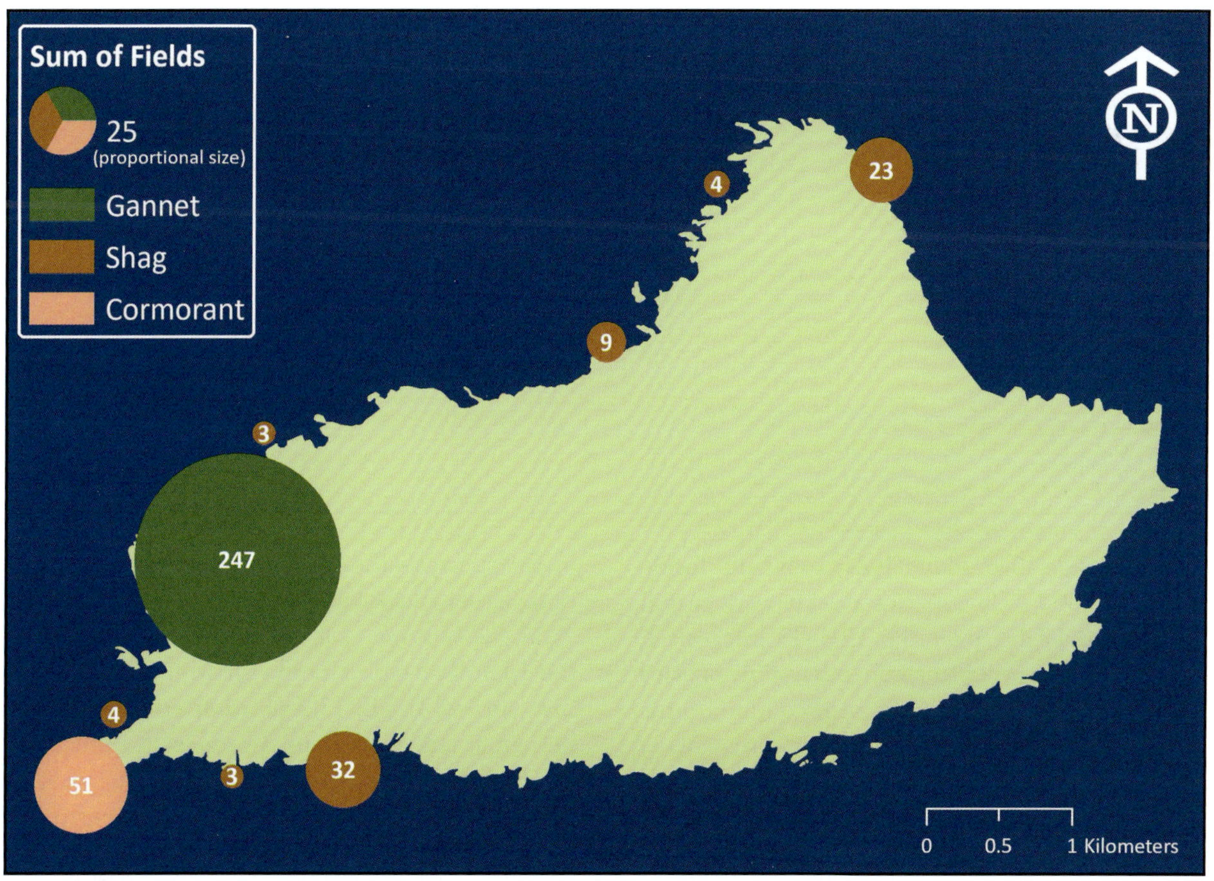

Fig. 3 Distribution and numerical abundance of the gannet, great cormorant and shag on Clare Island.

Table 1
Species of seabird, their clutch sizes and breeding season foraging zones.

Species	Clutch size	Breeding season feeding location
Fulmar	1	Pelagic
Gannet	1	Pelagic
Great cormorant	3–4	Neritic
Shag	3	Neritic
Great skua	2	Neritic
Common gull	3	Neritic
Lesser black-backed gull	3	Neritic
European herring gull	3	Neritic
Great black-backed gull	2–3	Neritic
Kittiwake	2	Neritic/Pelagic
Common guillemot	1	Neritic
Razorbill	1	Neritic
Black guillemot	1–2	Neritic
Puffin	1	Neritic/Pelagic

Table 2
Weights of male and female of 14 species of seabird breeding on Clare Island; weights mainly from Dunning (2008). Note separate weights for male and female razorbill, black guillemot and puffin are not given—so it is assumed for the purposes of the broad scale metric used here that the weights of male and female are the same. The total biomass was estimated by multiplying the combined average by the estimated number of individuals of each species at the colony in 1999 and 2015.

Species	Average weight male (g)	Average weight female (g)	Average weight combined (g)
Fulmar	649	577	613
Gannet	2932	3067	3000
Great cormorant	3420	2630	3025
Shag	1948	1598	1773
Great skua	1271	1409	1340
Common gull	432	348	390
Lesser black-backed gull	880	755	818
European herring gull	1147	1023	1085
Great black-backed gull	1829	1488	1659
Kittiwake	223	183	203
Common guillemot	1006	979	993
Razorbill	726	726	726
Black guillemot	381	381	381
Puffin	378	378	378

and Ruttledge 1945; Lloyd 1984) could not locate any colony and so this species must have departed from Clare Island over the 1911–45 interval. It is notable, moreover, that no other tern species breeds on the island though both common terns *Sterna hirundo* Linnaeus, sandwich terns *Sterna sandvicensis* Latham, and arctic terns are occasionally seen either around the harbour or from the ferry on a journey to and from the mainland.

Six species—the fulmar, common gull, cormorant, lesser black-backed gull, gannet and great skua *Stercorarius skua* (Brünnich)—have joined the island's seabird fauna since the time of the first survey. In all, 15 seabird species have been known to have bred on Clare Island since the original survey (Ussher 1912; this study). This gain of six species (66.7%; 75% if the departure of the arctic tern is considered) over the total listed

Table 3
Seabird totals (standard census units) counted on Clare Island in 1999 and 2015–17.

Species	1999	2015–17	Census unit	% Change
Fulmar	4029	2505	Apparently Occupied Site (AOS)	–38%
Gannet	3	352	Apparently Occupied Nest (AON)	+>100%
Great cormorant	64**	51	Apparently Occupied Nest (AON)	Unknown
Shag	93	78	Apparently Occupied Nest (AON)	–16.12%
Great skua	0	1	Apparently Occupied Territory (AOT)	+100%
Common gull	45	12	Apparently Occupied Nest/Territory (AON/AOT)	–73%
Lesser black-backed gull	14	51	Apparently Occupied Nest/Territory (AON/AOT)	+>100%
European herring gull	23	25	Apparently Occupied Nest/Territory (AON/AOT)	+8.7%
Great black-backed gull	24	46	Apparently Occupied Nest/Territory (AON/AOT)	+92%
Kittiwake	1785	840	Apparently Occupied Nest (AON)	–54%
Common guillemot	2280*	2168	Individuals	–9%
Razorbill	528*	618	Individuals	+13%
Black guillemot	42	98	Individuals	+>20%
Puffin	48	Not Counted	Individuals	Unknown

The estimates (*) of the total number of pairs are based on: (i) the 1999 census of puffin and black guillemot, and (ii) a conversion factor k of 1 individual on a ledge = 0.67 pair for common guillemot and razorbill (see Lloyd *et al.* 1991; Cussen *et al.* 1999).
** refers to individuals.

by Ussher (1912) represents a remarkable change in the composition of the seabird community of Clare Island and an exceptional increase in the species richness of the marine bird fauna (see also Chapter 7).

Lloyd (1984) summarised the known history of the colonisation of Clare Island by the fulmar, which first bred in Ireland in 1911 (Fisher 1966). Prospecting of the cliffs of the island was first detected in 1915 but breeding was not confirmed until 1943 (Fisher 1966). Lloyd (1984) is surely correct in asserting that 'breeding probably started some time just after 1935 because, by 1945, fulmars were to be found almost continuously along the western cliffs'.

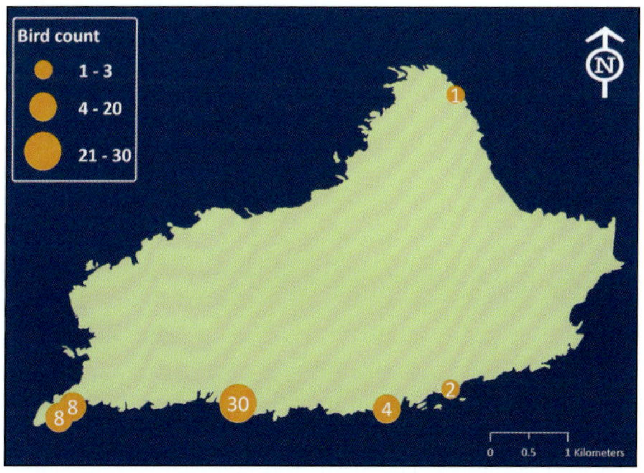

Fig. 4 Distribution and numerical abundance of the common gull on Clare Island in 1999 and 2015.

The gannet first bred on Clare Island in 1978 when one nest was occupied on the small stack off the north-east corner of the island (see Lloyd 1984).

Common gulls were not recorded by Ussher (1912), but five pairs were found by Barlee and Ruttledge (1945) at Kinatevdilla. This latter area was not surveyed in 1982 (Lloyd 1984), but a total of 45 nests were counted during the 1999 survey (Cussen *et al.* 1999) and while numerical abundance was not estimated in 2013, the species was seen in suitable breeding habitat on the coast at Craigmore—the site of the largest number of nests in 1999 (Cussen *et al.* 1999). However, in the 2015 survey breeding common gulls were recorded in much lower numbers and were confined to this single site.

The lesser black-backed gull was first recorded breeding in 1969 during 'Operation Seafarer' the first seabird census of Britain and Ireland (Cramp *et al.* 1974).

The cormorant was first recorded as a breeding species on Clare Island in 1999 when nests—and at least 64 individuals—were discovered on Kinatevdilla; the island appears to have been colonised after 1990, in which year no breeding birds were recorded (Cussen *et al.* 1999). Cormorants were also counted (n=40+) during a survey made by Richard Coombes in late April and early May 2000 (R.H. Coombes, pers. comm.). On 20 April 2016 a count of

44 apparently occupied nests was obtained. In 1985, during the Seabird Colony Register (Lloyd *et al.* 1991), a count of 50 occupied cormorant nests was made on Mweelaun, an island approximately 4km southeast of the Kinatevdilla colony. As this breeding site was deserted subsequent to this date it is quite probable that it is from here that the Kinatevdilla colony originated. The latest estimate is of 51 AONs counted on 5 May 2017 by Eoin McGreal.

A territorial pair of great skuas was first seen on the island in 2015 and breeding was confirmed in 2016. This follows its arrival in and the colonisation of Ireland after a large increase in the Scottish breeding population; Caher Island, some 9km to the south of Clare Island was one of its initial Irish breeding sites, in 2000, while Inishturk has hosted a breeding pair since 2006.

Changes in status up to 1999

Detailed comparisons with previous counts (principally those made in 1982 and 1990) must take account of the possible differences in survey methodology, the completeness of counts and the boundaries of the cliff sections counted in the different years. This information may not always be available, and counts may have to be taken at face value (with the proviso that unquantifiable errors may be involved).

Of the two surveys mentioned above, the 1982 survey by Lloyd (1984) covered all the island's breeding seabirds, while data available from the more recent counts by Delany and Liggett indicates that *Larus* gulls were not surveyed (Seabird Colony Register, unpublished). A simple comparison between these surveys for Clare Island would suggest that there have been major changes in the numbers of breeding seabirds since the previous surveys, including the addition of a new breeding species (great cormorant). However, some of these changes may be the result of underestimates for areas of cliff between the old Signal Tower on the west coast and the sea stack to the east of Knockmore in the earlier surveys.

Comparison between the 1990 and 1999 counts suggests major increases in common guillemot (up to 255% from 642 individuals), fulmar (up 112% from 1,898 AOS), razorbill (up 100% from 264 individuals), puffin (up 140% from 20 individuals) and gannet (up 50% from two AON), while kittiwake (up 4% from 1712 AON) show little change. A comparison with the 1982 survey for the remaining species also suggests major increases in lesser black-backed gull, up 1,300% (from one AON) and shag, up 365% (from 20 AON). Only 12 individual common gulls were recorded in May 1982 (A. Whilde in Lloyd 1984). For most of the above species the scale of increases may simply be due to underestimation in earlier surveys, as it is not clear to what extent (if any) boat-based counts were made in 1982.

In contrast, the 1999 counts suggest a 95% decrease in herring gull numbers and a 20% decrease in those of the great black-backed gull since the 1982 survey. Black guillemot also showed a slight decrease of 9% on the 1982 counts. The decrease in herring gull numbers would be consistent with the long-term decline seen in many Irish colonies in recent decades (Lloyd *et al.* 1991; Mitchell *et al.* 2004). The decrease in great black-backed gull numbers may have been due to competition for nest space at the main colony site with the recently established cormorant colony.

However, the possible breeding of the storm petrel *Hydrobates pelagicus* (Linnaeus) and the manx shearwater *Puffinus puffinus* (Brünnich) remains unconfirmed (see especially Lloyd 1984); recent surveys have not discovered evidence of the presence of either species on the island, but the required nocturnal visits to the cliff-top habitats on cloudy nights do not appear to have been attempted.

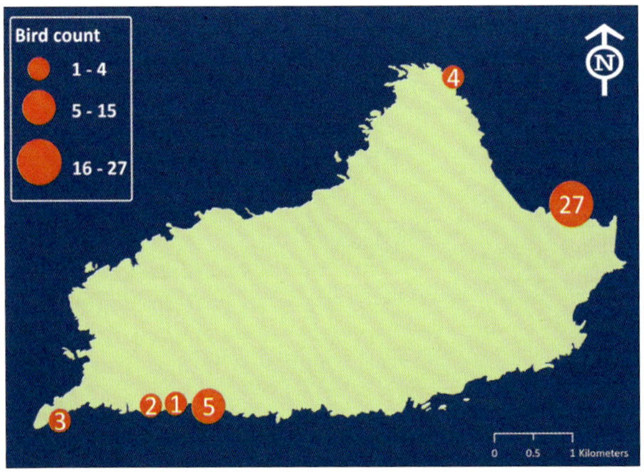

Fig. 5 Distribution and numerical abundance of the black guillemot on Clare Island in 1999.

Changes in status since 1999

As can be seen in Table 3, major decreases have occurred in the population size of the fulmar (–38%), shag (–16%), common gull (–73%) and kittiwake (–54%) over the 16-year interval since the 1999 survey. The number of common guillemots (individuals) showed a 9% decline but given the difficulties of censusing this species from the sea it is perhaps premature to define this as a definite trend. Other species including the great black-backed gull (+92%), lesser black-backed gull (+>100%), razorbill (+15%), herring gull (+8.7%) and—spectacularly—the gannet from three AON in 1999 to 267 AON in 2015 (McGreal unpublished; Newton *et al.* 2015a, 2015b). The cause (or causes) of this remarkable increase, which parallels a similar trend for other gannet colonies in Ireland is unknown (see Newton *et al.* 2015a). Likewise, the reasons why the white-headed gull colonies are expanding on Clare Island remain unexplained as the great black-backed gull feeds primarily on marine prey whereas the lesser black-backed gull obtains a substantial proportion of its diet in the terrestrial zone (e.g. Buckley 1990; Buckley and Kelly 1994; Crème 1995; Kelly *et al.* 2012; Kelly and Kelly in prep). The herring gull population may now have stabilised as it is a predator of the dog whelk *Nucella lapillus* and other Mollusca in the intertidal zone (Cussen and Myers 2002). However, it has also been increasingly seen in fields apparently feeding on terrestrial invertebrates (Kelly unpublished data).

Overall, however, the size of the seabird colony on Clare Island appears to have decreased between 1999 and 2015 by >2000 pairs or 25%, with greatest declines in the fulmar and kittiwake.

The decrease in the numerical abundance of the fulmar coincides with the trend detected elsewhere around the coasts of Britain and Ireland (Russell *et al.* 2015; JNCC 2015). Notably however, the scale of the decline on Clare Island appears to be double that detected among surveyed colonies around the British Isles (-38% compared to -18%; see JNCC 2015), though again caution needs to be exercised

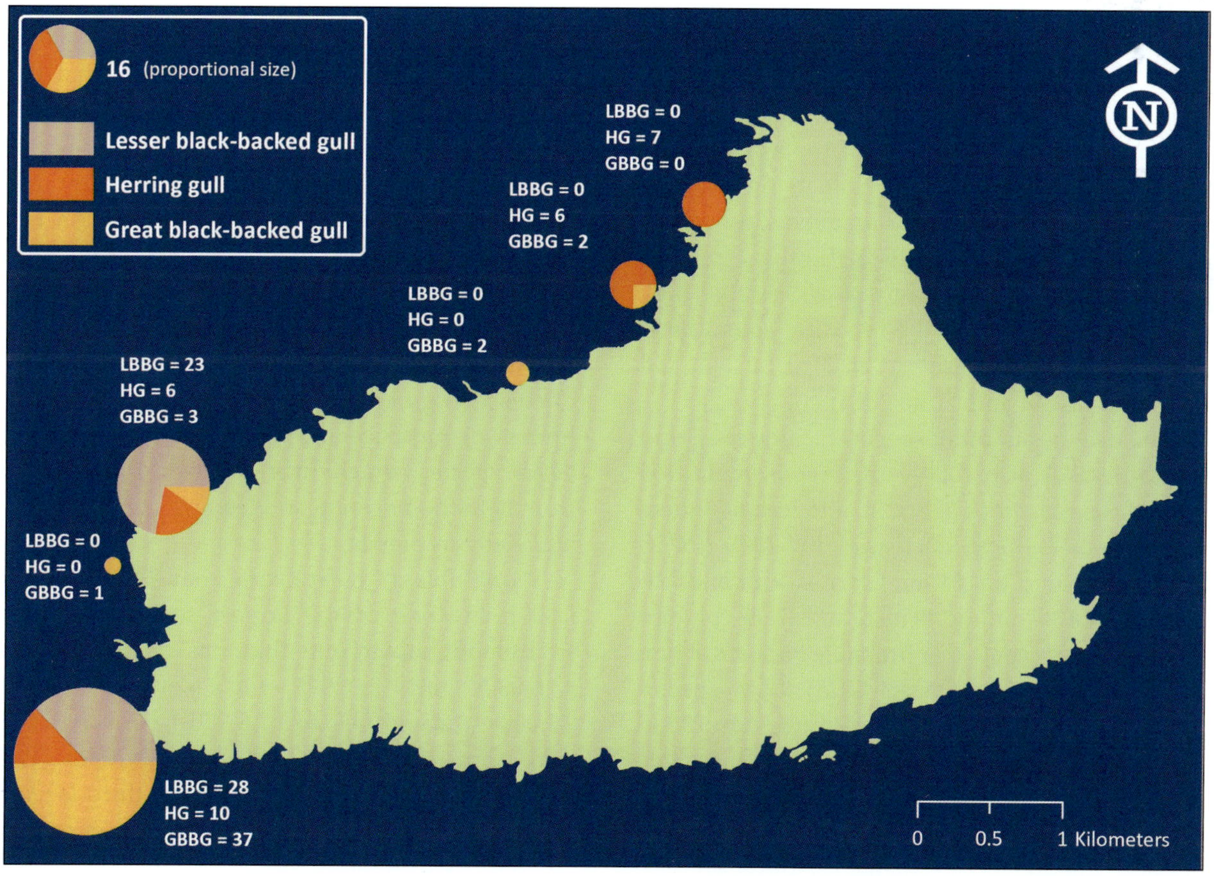

Fig. 6 Distribution and numerical abundance of the lesser black-backed gull, European herring gull and great black-backed gull on Clare Island.

in reading these data as exact values (there are, for example, no standard error estimates from the Clare Island surveys which in part, perforce, have to be conducted from the sea), but the trend is clear cut. The steep decline of the fulmar over much of North Mayo, where it first colonised Ireland in 1911 (Fisher 1966; Hutchinson 1989), is of special interest and concern, though it should be noted that the population on the nearby island of Inishturk has remained stable over the same interval. Moreover, since Seabird 2000 (Mitchell *et al.* 2004), the pattern of population fluctuation has been unexpectedly uneven; thus, while the large fulmar colony on Horn Head (Donegal) has undergone a marked decline, that on the Cliffs of Moher (Clare) has increased. The fulmar feeds mainly on sandeels (*Ammodytes*) during the breeding season (Cury *et al.* 2011). Unfortunately, there are no comparable published studies for fulmar colonies in Ireland and so it is not possible to identify its diet or, equally importantly, interpret any changes that may have occurred due to climate change or competition with fisheries (Cury *et al.* 2011). Ranging over the open ocean and indeed also in inshore waters, the fulmar, along with other seabird species, is exposed to numerous potential hazards, including contamination from oil, long-line fisheries, microplastics pollution and windfarms, while human disturbance and predation by introduced mammals, especially rats *Rattus* species, feral cats *Felis catus* Linnaeus and American mink *Neovison vison* Schreber may exert direct impacts on the breeding colony (e.g. Mitchell *et al.* 2004; Masden *et al.* 2010; Lusher *et al.* 2014; Ivar do Sol and Costa 2014).

The declines in other species (i.e. the shag and the kittiwake) also follow the trends identified around the coastlines of Britain and Ireland where shag numbers have decreased by 38% (JNCC 2015), compared to 16% on Clare Island, and kittiwake by 47% among British and Irish colonies generally, compared to a probable 54% reduction on Clare Island (see JNCC 2015). Both species prey on sandeels during the breeding season and the kittiwake appears to be relatively inflexible and therefore more vulnerable to a decrease in the availability of its preferred prey (*Ammodytes*) possibly due to climate change (Frederiksen *et al.* 2005, 2007). However, recent studies on the diet of kittiwakes on Rathlin (Antrim) and Lambay Islands (Dublin) (Chivers *et al.* 2012a) show

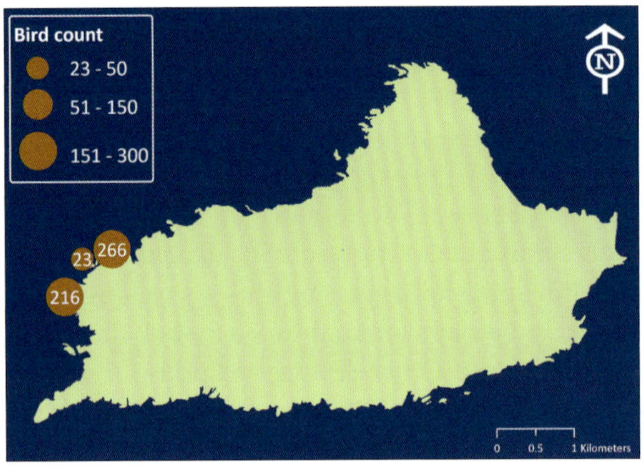

Fig. 7 Distribution and numerical abundance of the kittiwake on Clare Island in 2015.

sandeels were a major component in the latter but not the former colony, although on Rathlin, razorbills and common guillemots over the same interval were able to provision their chicks with *Ammodytes* (Chivers *et al.* 2012b). These studies emphasise the complexity of issues relating to food availability and how these may manifest themselves at local and wider scales (Bull *et al.* 2004; Chivers 2008).

The extent of the decline in the numbers of breeding shag on Clare Island is somewhat less than has been detected elsewhere in Britain and Ireland (-38%, JNCC 2015). This is a species that is known to experience high adult mortality during prolonged stormy weather (e.g. Harris and Wanless 1996) such as has occurred in Ireland in recent winters, e.g. January 2015 (Anonymous 2015).

The relatively massive decrease in the numbers of common gulls on Clare Island could be due to human disturbance and the increase in the larger *Larus* species which may have displaced them from both their nesting and feeding sites (see Kelly and Kelly in prep).

Nonetheless, it appears that Clare Island still holds one of the largest seabird colonies in Ireland, albeit one that has declined from *c.* 8,000+ pairs in 1999 and a possible total of 16,000+ individuals to *c.* 6,000+ pairs in 2015 and approximately 12,000+ individuals. The fulmar colony (n=2,505 AOS) is the third largest in Ireland and in total may number in excess of 5,000 individuals.

If these figures are correct, then some 52% of the seabirds breeding on Clare Island are pelagic and feed primarily in the offshore zone (Table 2).

The continental shelf is much closer to seabird colonies in North Mayo than those, for example, on the Cliffs of Moher, the Saltee Islands (Wexford) and Lambay Island (see Jessopp and Kelly 2012), which may explain the exceptionally high numbers of fulmars on this coastline (see Lloyd 1984; McGreal 2013).

There are also large numbers of breeding seabirds that feed in shallow inshore waters—the so-called neritic species—approximately 48% of the total on the island, if the kittiwake is included (Table 2). Casual observations including those made in June 2013 identified what appears to be a large front lying between Achill Sound and the western extremity of Clare Island. Large numbers of seabirds including gannets, which were continuously diving into this feature, were seen to be actively feeding in this area. The total biomass of seabirds (before the eggs hatch) on Clare Island may be as high as 10,000+kg (10.7 metric tonnes) (Table 3), which could lead to up to 3,000kg (three metric tonnes) of food being consumed by the colony every day (Furness 1978; Votier *et al.* 2004).

Unfortunately, nothing is known of the prey being taken by seabirds on Clare Island but other studies have shown that cormorants obtain a substantial proportion of their food from fresh water locations and, as mentioned, gulls (possibly excluding great black-backed) consume large quantities of terrestrial invertebrates (e.g. Crème 1995; Kelly *et al.* 2012).

Substantial changes in the numerical abundance of seabirds occurred on Clare Island between 1982 and 2015 (Cussen *et al.* 1999; Newton *et al.* 2015a, 2015b) with an overall loss of some 2,000+ pairs or 4,000+ individuals. More recently there has been a remarkable change in the status of the gannet; in 2004 there were three apparently occupied nests but between that year and 2014, the number had risen to an estimated 267 and a total count of 510 adults resulting in the fledging of some 142 young (Eoin McGreal in Newton *et al.* 2015a). This relatively massive increase is unlikely to have been achieved by incremental population growth and must have resulted, in part, from immigration. The most recent census conducted on 5 May 2017

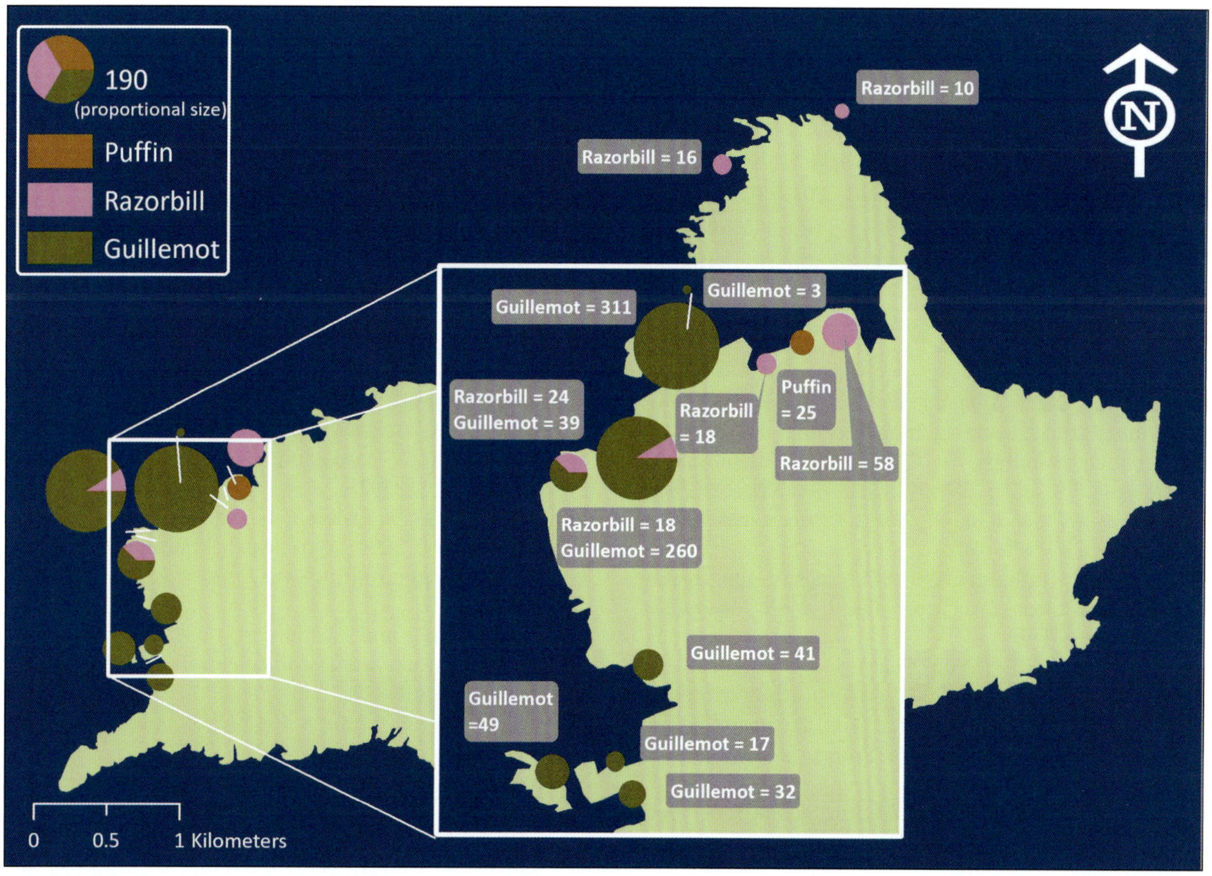

Fig. 8 Distribution and numerical abundance of the common guillemot, razorbill and puffin on Clare Island.

yielded an estimate of 352 AONs (Eoin McGreal, pers. comm.).

Little is known of the historical population strength of seabirds on Clare Island. At least one pair of golden eagles *Aquila chrysaetos* (Linnaeus) occupied an eyrie until around 1887 (Ussher 1912)—'in winter, when they had not sea-birds'—suggests that marine birds may have been an important component in their diet. Peregrine falcons *Falco peregrinus* Tunstall, which are resident on the island, would also take seabirds. More recently, there has been the remarkable colonisation of Northern Ireland and North Mayo by the great skua (Newton *et al.* 2002), which is now well established (e.g. Balmer *et al* 2013). This species is a notorious predator of seabirds (e.g. Votier *et al* 2004). Therefore, the confirmation of breeding of the great skua on Clare Island in 2016 by Eoin McGreal (unpublished data) may prove to be a major development in the history of the seabird fauna of Clare Island.

Man is also likely to have influenced the numbers of seabirds. Although there appears to be no mention of 'fowling' of seabirds by man on Clare Island (as occurred on the Faroe, Shetland and Blasket Islands (Kerry), for example) (Fisher 1952), the use of monofilament drift nets for catching salmon *Salmo salar* Linnaeus—now discontinued—was a not inconsiderable source of mortality among the auk species, especially razorbills and common guillemots (e.g. Whilde 1979; Buckley 1990). There may also have been costs and benefits for various species from the fisheries that developed in the region in the 1800s (see de Courcy Ireland 1981) and the North Mayo whaling stations established in the early twentieth century, which attracted 'clouds of gulls' (Fairley 1981). Herring fisheries attracted boats to Clare Island from as far away as Kinsale (Cork), but the arrival of modern trawling saw the effective closing of this activity in the late 1930s (Mac Cárthaigh 1999). The modern salmon farm along the east side of the island attracts small numbers of kittiwakes, fulmars and assorted tern and gull species. However, the food provisions for these salmon cages were stored in the harbour area and this resource attracted very large numbers of great black-backed gulls (>150) in the mid-2000s (as well as large numbers of hooded crows *Corvus cornix* Linnaeus) (see Chapter 2); great black-backed gull is a major predator of the eggs and young of seabirds but there are no data on what effect this spike in great black-backed gull numbers had on vulnerable species on Clare Island. The practice of this resource being available to gulls and crows was not evident in June 2013.

Acknowledgements

The boat survey team in June 1999 included the late Commandant Michael Hartnett; the boat survey conducted on 10 June 2015 was carried out with the assistance of National Parks and Wildlife Service colleagues Ger O'Donnell, Aongus Ó Donail and Dermot Breen.

REFERENCES

Anonymous. 2015 *Monthly Weather Bulletin* No. 344 January 2015 Glasnevin. Met Éireann.

Balmer, D.E., Gillings, S., Caffrey, B.J., Swann, R.L., Downie, I.S. and Fuller, R.J. 2013 *Bird Atlas 2007–11: the breeding and wintering birds of Britain and Ireland.* Thetford. BTO Books.

Barlee, J. and Ruttledge, R.F. 1945 Notes on the present status of birds on Clare Island. *Irish Naturalists' Journal* **8**, 311–3.

Brooke, M. de L. 2002 Seabird systematics and distribution: a review of current knowledge. In E.A. Schreiber and J. Burger (eds) *Biology of marine birds*, 57–85. London, Boca Raton, New York and Washington. CRC Press.

Buckley, N.J. 1990 Diet and feeding ecology of great black-backed gulls (*Larus marinus*) at a southern Irish breeding colony. *Journal of Zoology (London)* **222**, 363–73.

Buckley, N.J. and Kelly, T.C. 1994 Breeding biology of great black-backed gulls *Larus marinus* at a declining colony: Cape Clear Island, Co. Cork. *Irish Naturalists' Journal* **24**, 388–92.

Bull, J., Wanless, S., Elston, D.A., Daunt, F., Lewis, S. and Harris, M.P. 2004 Local scale variability in the diet of black-legged kittiwakes *Rissa tridactyla*. *Ardea* **92**, 43–52.

Burnham, K.P. and Anderson, D.R. 1984 Tests of compensatory versus additive mortality in mallards. *Ecology* **65**, 105–12.

BWP. *Birds of the Western Palaearctic Interactive*. BWPi Version 2.

Chivers, L.S. 2008 Breeding seabirds on Rathlin Island, County Antrim – a cause for concern. *Irish Birds* **8**, 359–64.

Chivers, L.S., Lundy, M.G., Colhoun, K., Newton, S.F. and Reid, N. 2012a Diet of black-legged kittiwakes

(*Rissa tridactyla*) feeding chicks at two Irish colonies highlights the importance of clupeids. *Bird Study* **59**, 363–7.

Chivers, L.S., Lundy, M.G. and Reid, N. 2012b Stable breeding despite variable feeding in two sympatric auks (Alcidae) species. *Bird Study* **59**, 67–73.

Clancy, C.F., O'Callaghan, M.J.A. and Kelly, T.C. 2006 A multiscale problem arising in a model of avian flu virus in a seabird colony. *Journal of Physics* (conference series) **55**, 45–54.

Coulson, J.C. 2002 Colonial breeding in seabirds. In E.A. Schreiber and J. Burger (eds) *Biology of marine birds*, 87–113. London, Boca Raton, New York and Washington. CRC Press.

Cramp, S., Bourne, W.R.P. and Saunders, D. 1974 *The Seabirds of Britain and Ireland*. London. Collins.

Crème, G. 1995 Aspects of the biology of an expanding population of lesser black-backed gulls *Larus fuscus* in Ireland. Ph.D. thesis. Cork. University College.

Croxall, J.P. 1987 (ed.) *Seabirds–feeding ecology and role in marine ecosystems*. Cambridge. Cambridge University Press.

Cury, P.M., Boyd, I.L., Bonhommeau, S., Anker-Nilssen, T., Crawford, R.M., Furness, R.W., Mills, J.A., Murphy, E.J., Osterblom, H., Paleczny, S., Platt, J.F., Roux, J-P., Shannon, L. and Sydeman, W.J. 2011 Global seabird response to forage fish depletion – one third for the birds. *Science* **334**, 1703–6.

Cussen, R., Kelly, T.C., Hartnett, M. and Walsh, P.M. 1999 Counts of breeding seabirds on Clare Island, Co. Mayo, 1999. Dublin. Report to Royal Irish Academy for the New Survey of Clare Island.

Cussen, R. and Myers, A. 2002 Dog whelk *Nucella lapillus* (Linnaeus), a significant prey item in the diet of herring gulls on Clare Island. In A. Myers (ed.) *New Survey of Clare Island*. Volume **3**, 151–7. Dublin. Royal Irish Academy.

Dai, L., Vorselan, D., Korolev, K.S. and Gore, J. 2012 Generic indicators for loss of resilience before a tipping point leading to population collapse. *Science* **336**, 1175–7.

de Courcey Ireland, J. 1981 *Ireland's Sea Fisheries - a history*. Dublin. The Glendale Press.

Dunning, J.B. 2008 *CRC handbook of avian biomasses*. Second edition. London, Boca Raton and New York. CRC Press.

Durant, J.M., Hjermann, D.O., Frederiksen, M., Charrassin, J.B., Le Maho., Y., Sabarros, P.S., Crawford, R.J.M. and Stenseth, N.C. 2009 Pros and cons of using seabirds as ecological indicators. *Climate Research* **39**, 115–29.

Fairley, J.S. 1981 *Irish whales and whaling*. Belfast. Blackstaff Press.

Fisher, J. 1952 *The fulmar*. New Naturalist. London. Collins.

Fisher, J. 1966 The fulmar population of Britain and Ireland, 1959. *Bird Study* **13**, 5–76.

Frederiksen, M.F., Wright, P.J., Harris, M., Mavor, R.A., Heubeck, M. and Wanless, S. 2005 Regional patterns of kittiwake *Rissa tridactyla* breeding success are related to variability in sandeel recruitment. *Marine Ecology Progress Series* **300**, 201–11.

Frederiksen, M.F., Mavor, R.A. and Wanless, S. 2007 Seabirds as environmental indicators: the advantages of combining data sets. *Marine Ecology Progress Series* **352**, 205–11.

Furness, R.W. 1978 Energy requirements of seabird communities: a bioenergetics model. *Journal of Animal Ecology* **47**, 39–53.

Furness, R.W. 2007 Responses of seabirds to depletion of food fish stocks. *Journal of Ornithology* **148**, 247–52.

Goldstein, D.L. 2002 Water and salt balance in seabirds. In E.A. Schreiber and J. Burger (eds) *Biology of marine birds*, 467–83. London, Boca Raton, New York and Washington. CRC Press.

Harris, M.P., Heubeck, M., Newell, M.A. and Wanless, S. 2015a The need for year-specific correction factors (k values) when converting counts of individual common guillemots *Uria aalge* to breeding pairs. *Bird Study* **62**, 276–9.

Harris, M.P., Newell, M.A. and Wanless, S. 2015b The use of k values to convert counts of individual razorbills *Alca torda* to breeding pairs. *Seabird* **28**, 30–6.

Harris, M.P. and Wanless, S. 1996 Differential responses of guillemot *Uria aalge* and shag *Phalacrocorax aristotelis* to a late winter wreck. *Bird Study* **43**, 220–30.

Hutchinson, C.D. 1989 *Birds in Ireland*. Calton. Poyser.

Ivar do Sol, J.A. and Costa, M.F. 2014 The present and future of microplastic pollution of the marine environment. *Environmental Pollution* **185**, 352–64.

Jessopp, M. and Kelly, T.C. 2012 Open sea. In R. Nairn and J. O'Halloran (eds) *Bird habitats in Ireland*, 181–95. Cork. Collins Press.

JNCC. 1999 *Seabird 2000 – census instructions*. Aberdeen. Joint Nature Conservation Committee.

JNCC. 2015 *Seabird population trends and causes of change: 1986–2014 report* (http://www.jncc.defra.gov.uk/page-3201). Aberdeen. Joint Nature Conservation Committee. Updated October 2015. Accessed 21 July 2016.

Kelly, M., Oxbrough, A. and Kelly, T.C. 2012 Some observations on the diet of adult common gulls (*Larus canus*) at a coastal breeding colony in County Kerry. *Irish Birds* **9**, 357–60.

Lloyd, C.S. 1984 The birds of Clare Island, Co. Mayo in June 1982. *Irish Naturalists' Journal* **21**, 212–6.

Lloyd, C.S., Tasker, M.L. and Partridge, K. 1991 *The status of seabirds in Britain and Ireland*. London. Poyser.

Lusher, A.L., Burke, A., O'Connor, I. and Officer, M. 2014 Microplastic pollution in the Northeast Atlantic: validated and opportunistic sampling. *Marine Pollution Bulletin* **88**, 325–33.

Mac Cárthaigh, C. 1999 Clare Island folklife. In C. Mac Cárthaigh and K. Whelan (eds) *New Survey of Clare Island*. Volume **1**, 41–72. Dublin. Royal Irish Academy.

Mackey, M., Ó Cadhla, O., Kelly, T.C., Aguilar de Soto, N. and Connolly, N. 2004 *Cetaceans and seabirds of Ireland's Atlantic margin*. Volume 1. Seabird distribution, density and abundance. Cork. Coastal and Marine Resource Centre.

Masden, E.A., Haydon, D.T., Fox, A.D. and Furness, R.W. 2010 Barriers to movement: modelling energetic costs

of avoiding marine windfarms amongst breeding seabirds. *Marine Pollution Bulletin* **60**, 1085–91.
McGreal, E. 2013 Probable breeding by Leach's storm petrel *Oceanodroma leucorrhea* at Bills Rocks, County Mayo; results of a census of its breeding birds. *Irish Birds* **9**, 636–8.
Mitchell, P.I., Newton, S.F., Ratcliffe, N. and Dunn, T.E. 2004 *Seabird populations of Britain and Ireland*. London. Poyser.
Neubauer, P., Jensen, O.F., Hutchings, J.A. and Baum, J.K. 2013 Resilience and recovery of overexploited marine populations. *Science* **340**, 347–9.
Newton, S.F. 2012 Sea cliffs, islands and rocky coasts. In R. Nairn and J. O'Halloran (eds) *Bird habitats in Ireland*, 138–51. Cork. Collins Press.
Newton, S.F., Harris, M.P. and Murray, S. 2015a Census of gannet *Morus bassanus* colonies in Ireland in 2013–2014. *Irish Birds* **10**, 215–20.
Newton, S.F., Lewis, L. and Trewby, M. 2015b Results of a breeding survey of important cliff-nesting seabird colonies in Ireland 2015 – with an interim analysis of population changes. Unpublished Report. Kilkcoole. Birdwatch Ireland.
Newton, S.F., Moralee, A. and Cabot, D.B. 2002 Great skua *Catharacta skua* breeding in Ireland. *Irish Birds* **7**, 129.
Nichols, J.D. 1991 Responses of North American duck populations to exploitation. In Perrins, C.M., Lebreton, J-D. and Hirons, G.J.M. (eds) *Bird population studies*, 498–525. Oxford Ornithology Series 1. Oxford. Oxford University Press.
Peron, G. 2013 Compensation and additivity of anthropogenic mortality: life history effects and review of methods. *Journal of Animal Ecology* **82**, 408–17.
Phillips, R.A., Gales, R., Baker, G.B., Double, M.C., Favero, M., Quintina, F., Tasker, M.L., Weimerskirch, H., Uhart, M. and Wolfaarrdt, A. 2016 The conservation status and priorities for albatrosses and large petrels. *Biological Conservation* **201**, 169–83.
Pollock, C.M. and Barton, C. 2008 *A gap analysis of Irish waters using the European Seabirds at Sea database*. Irish Wildlife Manuals Report. Dublin. National Parks and Wildlife Service.
Pollock, C.M., Reid, J.B., Webb, A. and Tasker, M.L. 1997 *The distribution of seabirds and cetaceans in the waters around Ireland*. Report No. 267. Aberdeen. Joint Nature Conservation Committee.
Rivelan, P., Barbaud, C., Inchausti, P. and Weimerskirch, H. 2010 Combined impacts of longline fisheries and climate change on the persistence of the Amsterdam albatross *Diomedea amsterdamensis*. *Ibis* **152**, 6–18.
Rolland, V., Weimerskirch, H. and Barbaud, C. 2010 Relative influence of fisheries and climate change on the demography of four albatrosses. *Global Change Biology* **16**, 1910–22.
Russell, D.J.F., Wanless, S., Collingham, Y.C. and Hamer, K.C. 2015 Predicting future European breeding distributions of British seabird species under climate change and unlimited no dispersal scenarios. *Diversity* **7**, 342–59.
Saether, B.E. and Bakke, O. 2000 Avian life history variation and contribution of demographic traits. *Ecology* **81**, 642–53.
Sibley, R.M., Barker, D., Hone, J. and Pagel, M. 2007 On the stability of populations of mammals, birds, fish and insects. *Ecology Letters* **10**, 970–6.
Ussher, R.J. 1912 A biological survey of Clare Island in the county of Mayo, Ireland and adjoining district. *Proceedings of the Royal Irish Academy* **31**, Sec. 11, Part 20.
Votier, S.C., Furness, R.W., Bearhop, S., Crane, J.E., Caldow, R.W.G., Catry, P., Ensor, K., Hamer, K.C., Hudson, A.V., Kalmbach, E., Klomp, N.I., Pfeiffer, S., Phillips, R.A., Prieto, I. and Thompson, D.R. 2004 Changes in fisheries discard rates and seabird communities. *Nature* **427**, 727–30.
Walsh, P.M., Halley, D.J., Harris, M.P., del Nevo, A., Sim, I.M.W. and Tasker, M.L. 1995 *Seabird monitoring handbook for Britain and Ireland*. Peterborough. JNCC/RSPB/ITE/ Seabird Group.
Whilde, A. 1979 Auks trapped in salmon drift nets. *Irish Birds* **1**: 370–6.
Whittow, G.C. 2002 Seabird reproductive physiology and energetics. In E.A. Schreiber and J. Burger (eds) *Biology of marine birds*, 409–37. London, Boca Raton, New York and Washington. CRC Press.
Wynne-Edwards, V.C. 1964 Oceanic birds. In A.L. Thomson (ed.) *A new dictionary of birds*. London. Nelson.

CHAPTER 4

ASPECTS OF THE BREEDING LAND BIRD COMMUNITY ON CLARE ISLAND

T.C. Kelly, U. Keating and E. McGreal

ABSTRACT

The first survey of Clare Island's birds (1908–11) and subsequent surveys provided little data on the numerical abundance of breeding land bird species. The approach then, and subsequently, focused on species richness and this has been typical of most studies of the breeding land birds of Irish islands. The aim of the present study, therefore, based on data collected on 23 walked transects carried out in mid-June 2013, was to provide a preliminary estimate of the numerical abundance of the breeding land bird species on Clare Island. A total of 28 species (including pheasant *Phasanius colchicus* Linnaeus and common kestrel *Falco tinnunculus* Linnaeus) was recorded (but only four were migrants). The island land bird community was estimated to comprise >8,000 individuals and was dominated by open-habitat specialist and generalist species, e.g. meadow pipit *Anthus pratensis* (Linnaeus), northern wheatear *Oenanthe oenanthe* (Linnaeus), wren *Troglodytes troglodytes* (Linnaeus), dunnock *Prunella modularis* (Linnaeus) and blackbird *Turdus merula* Linnaeus. Despite the presence of some woodland, many hole-nesting species (e.g. the family Paridae) were absent, as was the goldcrest *Regulus regulus* (Linnaeus) and treecreeper *Certhia familiaris* Linnaeus. Only two *Carduelis* and one *Emberiza* species were breeding. Four crow species breed on Clare Island, but the rook *Corvus frugilegus* Linnaeus is absent. The land bird community appears to be unique in Ireland due to the exceptionally high numbers of wheatear breeding there. The breeding land bird community is dominated by species with a clutch size of four eggs, and those that produce two broods per annum. In terms of trophic relationships, predators of invertebrates accounted for over 95% of the island's total number of land birds. Herbivorous species, by contrast, were remarkably scarce and only amounted to <4% of the total. The estimated total biomass of land birds was >376.7kg or >24kg per km^2.

Introduction

Most studies of the birds breeding on the offshore islands of Ireland have focused on compiling species lists (for reviews see Hutchinson 1989; Kelly *et al.* 2014). These data are useful in Island Biogeography Theory (IBT)-type studies because the species richness of individual islands can be related to key variables such as island area

This area is just (<500m) to the west of the harbour beach. It contains a number of households and is the centre of land bird biodiversity on Clare Island. There are numerous gardens, a football pitch and several small marshes. There is some Foliage Height Diversity (FHD) although the tree layer is sparse. The marsh, which includes a ditch, lies to the north, west and south of the football pitch. This is where the sedge warblers *Acrocephalus schoenobaenus* (Linnaeus) breed – as they did in the time of the first (1909–1911) survey. Moorhens *Gallinula chloropus* (Linnaeus) have been seen here and reed buntings *Emberiza schoeniclus* (Linnaeus) also breed nearby. Photos: T. Kelly

and distance from the mainland (e.g. Reed 1980, 1981; Berry 2009). However, in other respects, and on its own, species richness is limited in what information it can convey about the avian communities occupying different islands. At a minimum, avian community ecology requires quantitative information about the numbers of breeding birds present on each island. The island bird community will then be defined by the relative abundance of each of the component species. Such quantitative data in turn will allow estimates to be made of gamma (total diversity i.e. species richness over all sampling units), alpha (average diversity over all sampling units), and beta (ratio of gamma and alpha diversity) bird species diversity (see Jurasinski *et al.* 2009 for an authoritative review of species diversity concepts), and this information can be used when comparisons are being made between the avian communities of different islands or indeed with those of different habitat types on the adjacent mainland.

However, estimating the numerical abundance of every land bird species (we define a land bird as any species that obtains most, but not necessarily all, of its food from terrestrial habitats) nesting on any island is not a trivial task and there are several well-known methods and constraints (e.g. Bibby *et al.* 1992, 2000). In addition to ambient weather conditions, the timing of the survey in relation to the annual breeding cycle and differences in the detectability of individual species are among the most important variables that will influence the reliability of field surveys attempting to estimate the numerical abundances of nesting birds in different terrestrial habitats (e.g. Burnham and Anderson 1984; Bibby *et al.* 2000; Alldredge *et al.* 2007; Buckland *et al.* 2001). While breeding bird numbers were often estimated in the past using a dedicated territory mapping technique (e.g. Lysaght 1989), this labour-intensive method, which required a minimum of ten independent visits to a relatively small survey area (Bibby *et al.* 1992), is not practicable to employ on most remote offshore islands. Nowadays, the territory mapping technique, has been largely replaced by the line transect (or point count census) sampling methodology (e.g. Bibby *et al.* 2000; Buckland *et al.* 2001; Gottschalk and Huettmann 2010). This latter technique, subject to meeting some key assumptions (e.g. that the observer accurately estimates the perpendicular distance between the transect line and the detected bird), enables the probability of detecting a species at different perpendicular distances to be calculated. Knowledge of the probabilities of detection can enable densities of species to be calculated within surveyed areas, subject to appropriate corrections being made to account for differences in the detectability of the various species encountered in the survey. The overall effect of applying this technique in the field, and the analysis of the data using the program Distance, is that an estimate of density ± an upper and lower bound of the error can be computed (Buckland *et al.* 2001; Gottschalk and Huettmann 2010; Thomas *et al.* 2013).

The aim of this study therefore was (i) to apply the transect method with distance bands and (ii) to use the program Distance to estimate the densities of breeding land birds on Clare Island, while accounting for differences in the detectability among species. This information is particularly relevant in island biogeography studies, as there has been a historical focus on investigating island species richness, while bird densities have been studied in less detail. Finally, it was our objective to establish a profile of the biological traits of the terrestrial bird species breeding on Clare Island.

Materials and methods

Dates and times of transect counts

All the fieldwork was conducted between 16 and 18 June 2013. Most transects were undertaken between 07.30 and 18.30hrs, but one was commenced at 21.45 and completed at 22.15hrs. This single set of transect counts would correspond in timing to the 'late season' interval in breeding bird surveys conducted in Britain and Ireland. However, the spring and early summer of 2013 were exceptionally cool in Ireland (Anonymous 2013a, 2013b, 2013c) and Britain. For example, in Ireland April 2013 saw low temperatures everywhere with recorded values in north Mayo being −1.5°C to −1.0°C below the Long-Term Average (LTA) and rainfall was between 125 and 150% above the LTA (Anonymous 2013a). Likewise, in May ambient air temperatures were again −1.5°C to −1.0°C below the LTA (especially in north Mayo) and rainfall was again above normal (Anonymous 2013b).

Data provided by the British Trust for Ornithology (BTO) in their *enews* bulletin showed that 'cold spring delays breeding birds' and 'migrant birds were late arriving'. 'Many early season nest box checks yielded only empty nest boxes and both resident and migrant birds appeared to have delayed their breeding' (Leuen Evans, BTO: info@bto-enews.org; received 19 July 2013).

By contrast, rainfall in June 2013 was 25 to 50% below normal in north Mayo and temperatures were >0.5° C above the LTA (Anonymous 2013c). Weather conditions during the transect counts were ideal: mostly cloudless skies and very light variable or north east winds (effectively calm). Ambient air temperatures were not measured but it was palpably warm on most days. While there was one short interval of light mist around midday on 17 June visibility was mostly excellent and these conditions were optimal for aural reception and identification of bird calls.

We are confident that in terms of the timing of the survey and the ambient conditions we optimised the probability of quantifying the numbers of breeding land bird species in June 2013. Very few fledged young were seen, and most breeding was confirmed by adult birds carrying food items.

Positions of transect counts

Transects were used to sample bird communities on Clare Island. The positions of the 23 transects walked over the 16 to 18 June 2013 interval are shown in Fig. 1. The positions of the transects were selected to avoid most areas of fenced-off privately owned agricultural holdings; there are large areas of commonage on Clare Island and most of the fenced-off areas can be readily surveyed from metalled roads or public boreens. The location of transects ensured that all the major habitats on the island were surveyed. However, riparian biota as well as those on the vegetated parts of the cliffs were not sampled due primarily to time constraints.

Field methods

In total, 19.7% (315ha) of Clare Island (1,598ha) was surveyed in this study. Birds were recorded 100m either side of the same experienced observer in two distance bands (0–50m, 50–100m) along defined transect routes that were perceived to

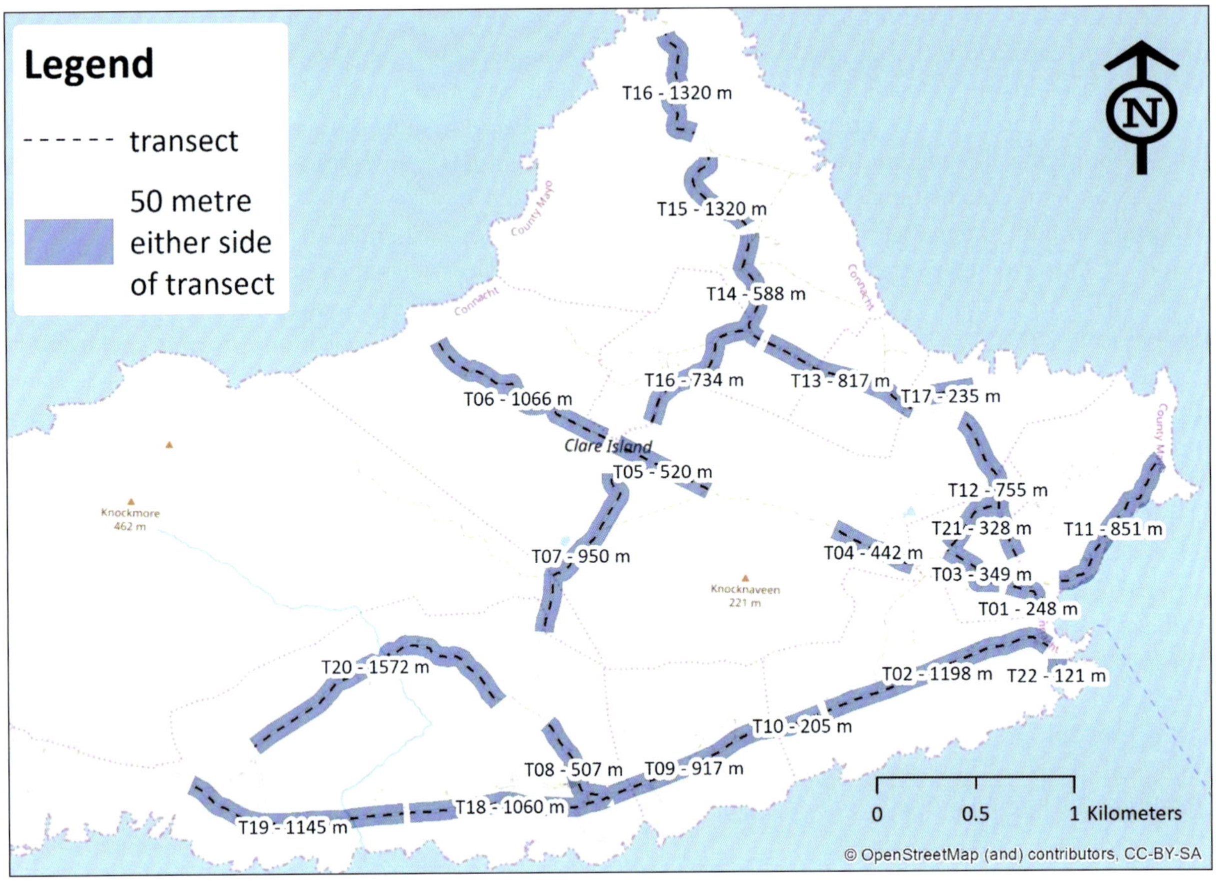

Fig. 1 Positions of 23 transects surveyed on Clare Island on 16 to 18 June 2013.

cover a representative subsample of the entire island. A total of 23 transects were surveyed (Fig. 1) in ideal conditions, that is, dry warm weather with light winds (<Beaufort Force 3) in June 2013. Visual registrations were made with 10x42 binoculars. The way points of the starting and end points of each of the transects were made with a Garmin GPS hand-held device. Transects were of variable length (mean 686m ± SE 78m). The maximum transect length was 1,572m and the minimum 121m. All sighted birds or those which were aurally detected were assigned to one of the two distance bands (0–50m, 50–100m) and these data were then analysed using the program Distance.

Distance 6.2 Release 1 (Thomas *et al.* 2010) was used to convert species count data within distance bands to density data through the generation of species detection functions. As it is recommended that a minimum of 60 registrations are required to generate reliable detection functions for species within a habitat (Buckland *et al.* 2001), detection functions were generated for all individual species with over 60 registrations. All other species were classified into two detection groups according to their perceived level of detectability, that is, small inconspicuous birds (commonly identified by their calls and designated as I), and large conspicuous birds (often identified by sight and designated as C) (Table 1). Species with less than 60 registrations were assigned to one of either of these groups based on the judgement of the surveyor after completing the census (Table 1). Following the generation of a detection function for both species groups, these were subsequently applied to individual species within their appropriate groups to get an estimate of species densities of all those with less than 60 registrations (see Fig. 2 a–d).

Table 1

The 28 land bird species detected in 23 transects walked over 16 to 18 June 2013 together with their mean densities per hectare ± SE as well as their biological traits based on the literature cited in the text and the detection group to which they were assigned for deriving species densities, accounting for differences in the probabilities of detection (C = conspicuous species group, I = inconspicuous species group; E = sufficient registrations to enable the compilation of an individual curve). An asterisk indicates migrant species.

Species	Density per ha	± SE	Clutch size	Brood size	Diet	Habitat association	Detection group
Blackbird	0.318	0.0037	3 eggs	2 broods	Invertebrate	Non-open	C
Chaffinch	0.032	0.0011	4 eggs	1 brood	Invertebrate	Non-open	I
Chough	0.041	0.0013	4 eggs	1 brood	Invertebrate	Non-open	C
Dunnock	0.246	0.0031	4 eggs	2 broods	Invertebrate	Non-open	I
Hooded crow	0.073	0.0015	3 eggs	1 brood	Invertebrate	Non-open	C
House sparrow	0.016	0.0007	4 eggs	2 broods	Invertebrate	Non-open	I
Common kestrel	0.006	0.0002	4 eggs	1 brood	Vertebrate	Non-open	C
Lesser redpoll	0.016	0.0007	4 eggs	2 broods	Herbivore	Non-open	I
Linnet	0.159	0.0020	4 eggs	2 broods	Herbivore	Non-open	I
Magpie	0.019	0.0005	4 eggs	1 brood	Invertebrate	Non-open	C
Meadow pipit	0.980	0.0077	4 eggs	2 broods	Invertebrate	Open-habitat	E
Pheasant	0.006	0.0002	8 eggs	1 brood	Omnivore	Non-open	C
Pied wagtail	0.103	0.0014	5 eggs	2 broods	Invertebrate	Non-open	I
Raven	0.013	0.0004	4 eggs	1 brood	Invertebrate; vertebrate	Non-open	C
Reed bunting	0.040	0.0009	4 eggs	2 broods	Invertebrate	Non-open	I
Robin	0.183	0.0022	4 eggs	2 broods	Invertebrate	Non-open	I
Rock dove	0.010	0.0002	2 eggs	2 broods	Herbivore	Non-open	C
Rock pipit	0.087	0.0022	4 eggs	2 broods	Invertebrate	Non-open	I
Sedge warbler*	0.032	0.0011	5 eggs	1 brood	Invertebrate	Non-open	I
Skylark	0.095	0.0012	3 eggs	2 broods	Invertebrate	Open-habitat	C
Song thrush	0.080	0.0015	4 eggs	2 broods	Invertebrate	Non-open	I
Common starling	0.321	0.0076	4 eggs	1 brood	Invertebrate	Non-open	E
Stonechat	0.199	0.0023	5 eggs	2 broods	Invertebrate	Non-open	I
Barn swallow*	0.326	0.0046	4 eggs	2 broods	Invertebrate	Non-open	I
Tree sparrow	0.111	0.0034	5 eggs	2 broods	Invertebrate	Non-open	I
Northern wheatear*	1.124	0.0097	5 eggs	1 brood	Invertebrate	Open-habitat	E
Willow warbler*	0.087	0.0014	5 eggs	1 brood	Invertebrate	Non-open	I
Wren	0.286	0.0034	5 eggs	2 broods	Invertebrate	Non-open	I

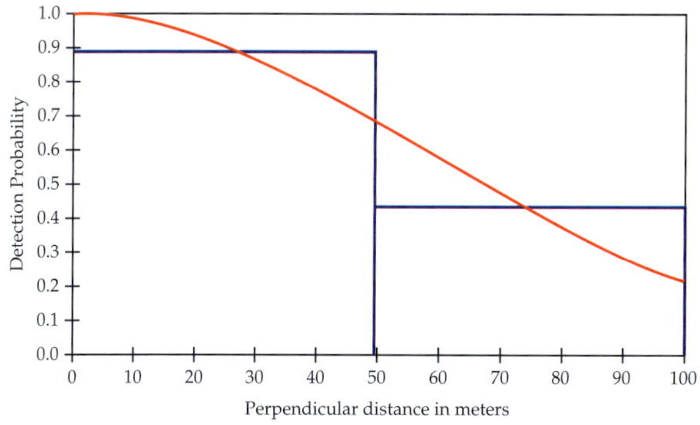

Fig. 2 a Detection curve for common starling.

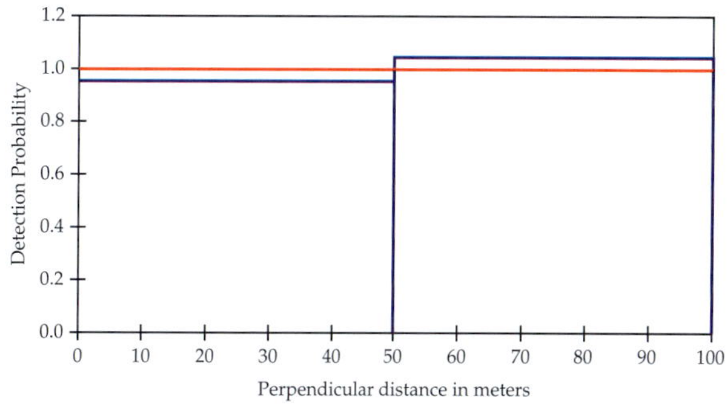

Fig. 2 b Detection curve for conspicuous species (C) with <60 registrations (100% probability of detection).

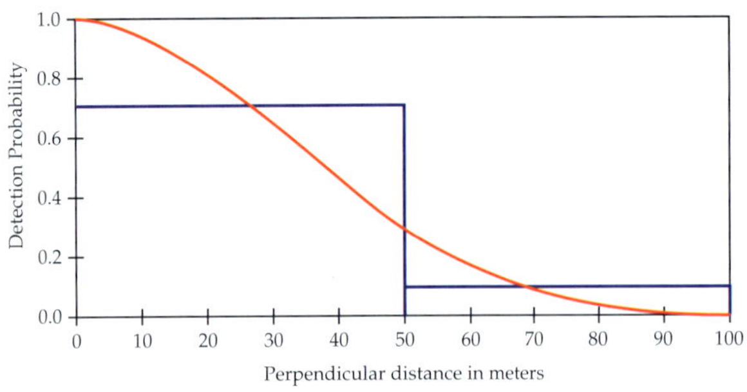

Fig. 2 c Detection curve for northern wheatear.

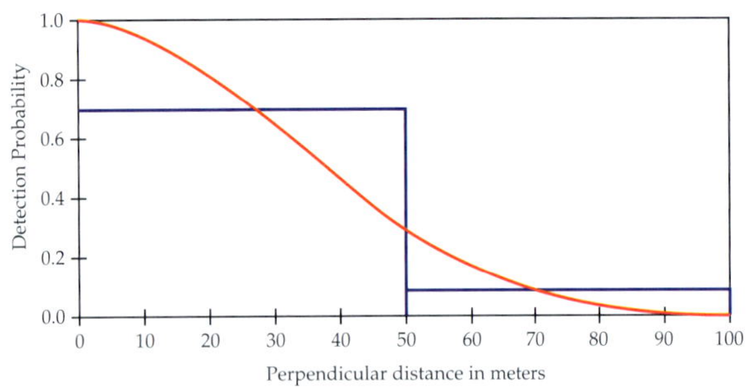

Fig. 2 d Detection curve for inconspicuous species (I) with <60 registrations.

The total abundance of each species was computed per transect area using density estimates derived for each transect. Species densities over the whole island were generated by calculating the total numbers of each species within the survey area of each transect (100m either side of the transect line), accounting for differences in detectability. The abundance of each bird species within the total surveyed area was calculated by adding together the abundances of species across all surveyed transects. The mean density of all bird species, per hectare of island, was calculated by dividing the species abundance within the total surveyed area by the total number of hectares surveyed. This value was multiplied by the total island area to get an estimate of the abundance of all species over the whole island. Standard errors were estimated by calculating the standard deviation of species abundance across transects and dividing by the square root of the number of transects (n=23). This value was divided by the area surveyed and subsequently multiplied by total island area.

Biological traits of breeding land bird species

Species were grouped based on their migratory status (migrant, resident), clutch size, number of broods and principal diet (i.e. the main food types fed to the chicks during the breeding season) (based on data in BWP*i* V2; Holland *et al.* 2006; Moorcroft *et al.* 2006). The broad habitat classification of land birds in Ireland followed O'Connell *et al.* (2012) and O'Halloran *et al.* (2012). In the case of clutch and brood sizes we used the minimum values given by Ferguson-Lees *et al.* (2011) since the reproductive output of bird species on islands tends to be lower than that recorded on adjacent mainland sites (e.g. Blondel 1985, 2000; Blondel *et al.* 1992; Smiddy *et al.* 1995; Fennessy 2001; Kelleher and O'Halloran 2006). A detailed list of species and their associated migratory, and dietary traits is given in Table 1.

The relative densities of these groups were compared. Using species density data and information on the mass (weight in grams) of individual species (Dunning 2007), the biomass of the bird community of Clare Island was calculated. Mean biomass was calculated by adding the mean masses of the male and female of each species (where known) and dividing by two. This figure was then multiplied by the estimated total number of individuals of each species, data for which was compiled as explained above.

Data on other land bird communities in Ireland was obtained *inter alia* from Lysaght (1989), Bracken and Bolger (2006), Wilson *et al.* (2006), Bracken *et al.* (2008), Sweeney *et al.* (2010), Bracken and Smiddy (2012), Cummins *et al.* (2012), Nairn and O'Halloran (2012), Sweeney *et al.* (2012), Wilson *et al.* (2012), Graham *et al.* (2015) and references therein.

Results

Species richness

A total of 28 species of land bird was found to be breeding on Clare Island in June 2013 (Table 1). Many of the resident species are habitat generalists, e.g. blackbird *Turdus merula* Linnaeus, song thrush *Turdus philomelos* C.L. Brehm, wren *Troglodytes troglodytes* (Linnaeus) and robin *Erithacus rubecula* (Linnaeus) (Nairn and O'Halloran 2012) and are widely distributed over the whole island of Ireland (Hutchinson 1989; Balmer *et al.* 2013). Notably, however, woodland specialists such as all the species of Paridae, long-tailed tit *Aegithalos caudatus* (Linnaeus), treecreeper *Certhia familiaris* Linnaeus and goldcrest *Regulus regulus* (Linnaeus) were absent (although see Winters (2006) in relation to goldcrest). Likewise, the endemic Irish jay *Garrulus glandarius hibernicus* (Witherby and Hartert) has not been recorded, though the Corvidae as a family are well represented on the island (Table 1). Only one species in the family Columbidae—doves and pigeons—is known to breed on Clare Island (i.e. the rock dove *Columba livia* J.F. Gmelin). The finch group is also depauperate with only the chaffinch *Fringilla coelebs* Linnaeus—a recent arrival—being present. The bullfinch *Pyrrhula pyrrhula* (Linnaeus) has not been recorded on the island and the genus *Carduelis* is represented by only two (i.e. linnet *C. cannabina* (Linnaeus) and lesser redpoll *C. cabaret* (P.L.S. Müller)) of the six species that breed in Ireland (Smiddy 2010). The reed bunting *Emberiza schoeniclus* (Linnaeus) is now the only *Emberiza* species breeding on the island.

Among the migrants, only two species of warbler breed (willow warbler *Phylloscopus trochilus* (Linnaeus) and sedge warbler *Acrocephalus schoenobaenus* (Linnaeus)). Remarkably, there was no *Sylvia* or *Locustella* warbler species breeding on Clare Island in 2013. The barn swallow *Hirundo rustica* Linnaeus is the sole Hirundinidae species on Clare Island (Table 1). Finally, the northern wheatear *Oenanthe oenanthe* (Linnaeus), as many surveyors remarked in the past, is a prominent species among the island's breeding avifauna.

Habitat associations

Breeding bird species were classified as either belonging to open or non-open habitats. Open habitat species comprised the northern wheatear, skylark *Alauda arvensis* Linnaeus and meadow pipit *Anthus pratensis* (Linnaeus) and together these amounted to 10.7% of the total land bird fauna. Therefore, those occupying the non-open habitats form the great majority of species breeding on the island (Fig. 3).

Breeding season diets

The diets of the breeding land birds on Clare Island (based on studies cited in BWP*i* V2) are summarised in Fig. 4, from which it can be seen the overwhelming majority (90+%) prey on invertebrates with which they provision their young. Predators of vertebrates are relatively rare as are omnivores (Fig. 4). The rock dove, linnet and lesser redpoll are the only herbivorous species breeding on Clare Island.

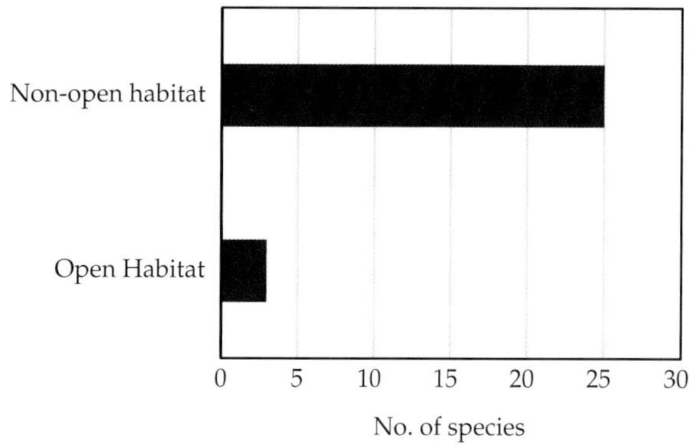

Fig. 3 Habitat associations of land bird species breeding on Clare Island (data mainly from Nairn and O'Halloran (2012); see also references cited in Methods).

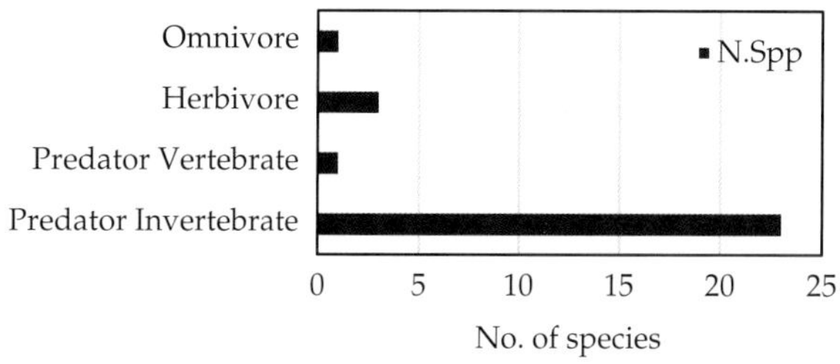

Fig. 4 Main food types fed to chicks of land bird species on Clare Island (based on BWP*i* V2).

Clutch and brood sizes

The range of clutch sizes of the land bird species breeding on Clare Island, as compiled from the literature (see Methods) is shown in Fig. 5. It is evident that the 'median' number of eggs is four, but it ranges from two to eight—though in the latter case the species is the pheasant *Phasanius colchicus* Linnaeus, which is not native to Ireland and is placed in the C1 category (Smiddy 2010); for an explanation see Parkin and Knox (2010). It should be noted again that we chose the lowest number of eggs laid by the different species (see Methods). Thus, the blackbird, hooded crow *Corvus cornix* Linnaeus and skylark all have clutch sizes of three (Ferguson-Lees *et al.* 2011), whereas the dunnock *Prunella modularis* (Linnaeus), common starling *Sturnus vulgaris* Linnaeus and barn swallow all have clutch sizes of four. The northern wheatear, willow warbler and sedge warbler—all migrant species—each have clutch sizes of five.

The known brood sizes (BWP*i* V2; Ferguson-Lees *et al.* 2011) are shown in Fig. 6. While 11 species (39.3%) have a single brood, a majority

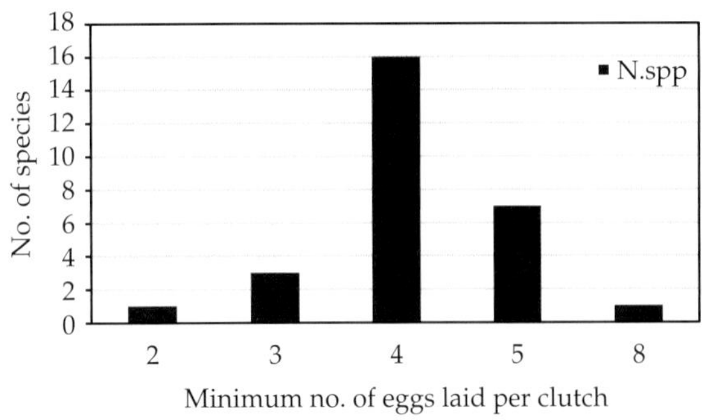

Fig. 5 Distribution of clutch sizes of bird species breeding on Clare Island (data from Ferguson-Lees *et al.* 2011 and BWP*i* V2).

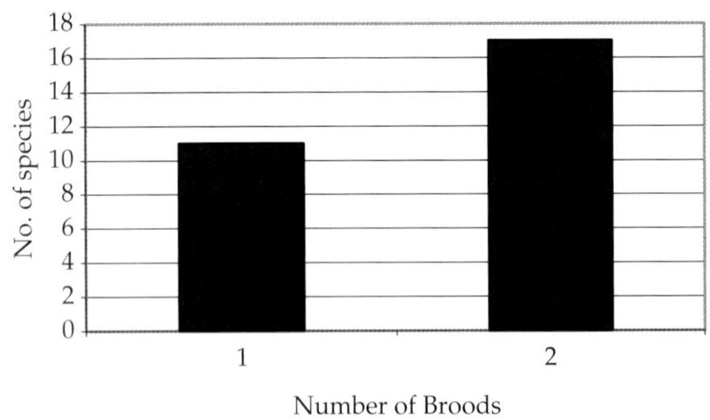

Fig. 6 Number of broods per year of 28 land bird species breeding on Clare Island (data from Ferguson-Lees *et al.* 2011 and BWP*i* V2).

(n=17; 60.7%) of species are double brooded. Interestingly, all the migratory species except the swallow are single brooded.

Resident versus migratory species

Only four (14.3%) of the 28 land bird species breeding on Clare Island are migrants (Fig. 7). The migratory species are the barn swallow, northern wheatear, willow warbler and sedge warbler.

Numerical abundance

The estimated numerical abundance of the breeding land bird species in June 2013 is shown in Fig. 8. Based on these estimates, the two most abundant species on the island are those occupying 'open habitats'—the northern wheatear (>1,700 individuals) and meadow pipit (>1,500 individuals) respectively (Fig. 8). However, the other 'open habitat' species, the skylark, is much less abundant (<200 individuals). It is notable that the tree sparrow *Passer montanus* (Linnaeus) is considerably more abundant than the house sparrow *Passer domesticus* (Linnaeus) (Fig. 8).

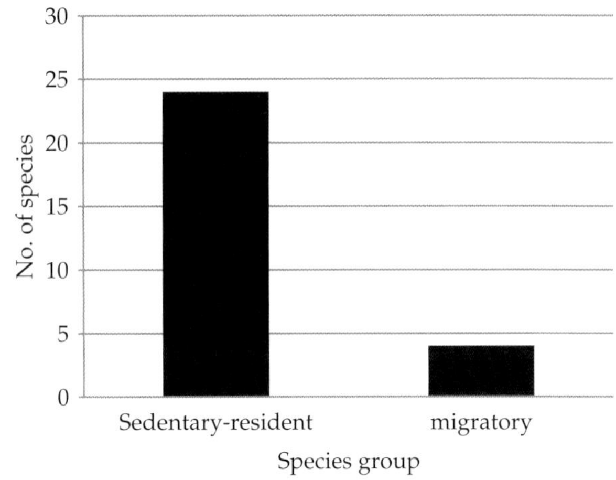

Fig. 7 Number of sedentary and migratory land bird species breeding on Clare Island in 2013 (classification based on Hutchinson (1989) and Wernham *et al.* (2002)).

The total number of land birds on Clare Island is >500 per km^2—a figure that lies between the estimated density of 635 individuals per km^2 in fen habitats and 353 per km^2 on raised bog—quoted by Bracken *et al.* (2008) (see also Bracken and Smiddy 2012). However, these authors report noticeably lower densities of 252 birds per km^2 on Atlantic blanket bog and 282 per km^2 on montane blanket bog respectively. Therefore, despite the large areas of blanket bog on Clare Island, the overall density of land birds is noticeably higher than might be expected. The wide presence of dry-stone walls, and in some parts hedges, could be responsible for the higher densities on Clare Island, as elevated numbers of passerine birds can utilize hedgerows and this habitat feature is often not present in bogs.

Species biomass

The biomass of the land birds breeding on Clare Island are shown in Fig. 9. These data are based on (a) the estimated total abundance of each species, as explained in the Methods, and (b) multiplied by the mean mass (in grams) of each taxon. Overall the biomass of land bird populations on Clare Island amount to approximately 376.7kg.

The hooded crow (>66kg) emerges as the heaviest species on Clare Island, followed by the blackbird (>57kg), northern wheatear (>54kg) and common starling (>44kg) (Fig. 9). The chough *Pyrrhocorax pyrrhocorax* (Linnaeus), rock dove and raven *Corvus corax* Linnaeus occur in relatively low numbers on the island, and hence, have a

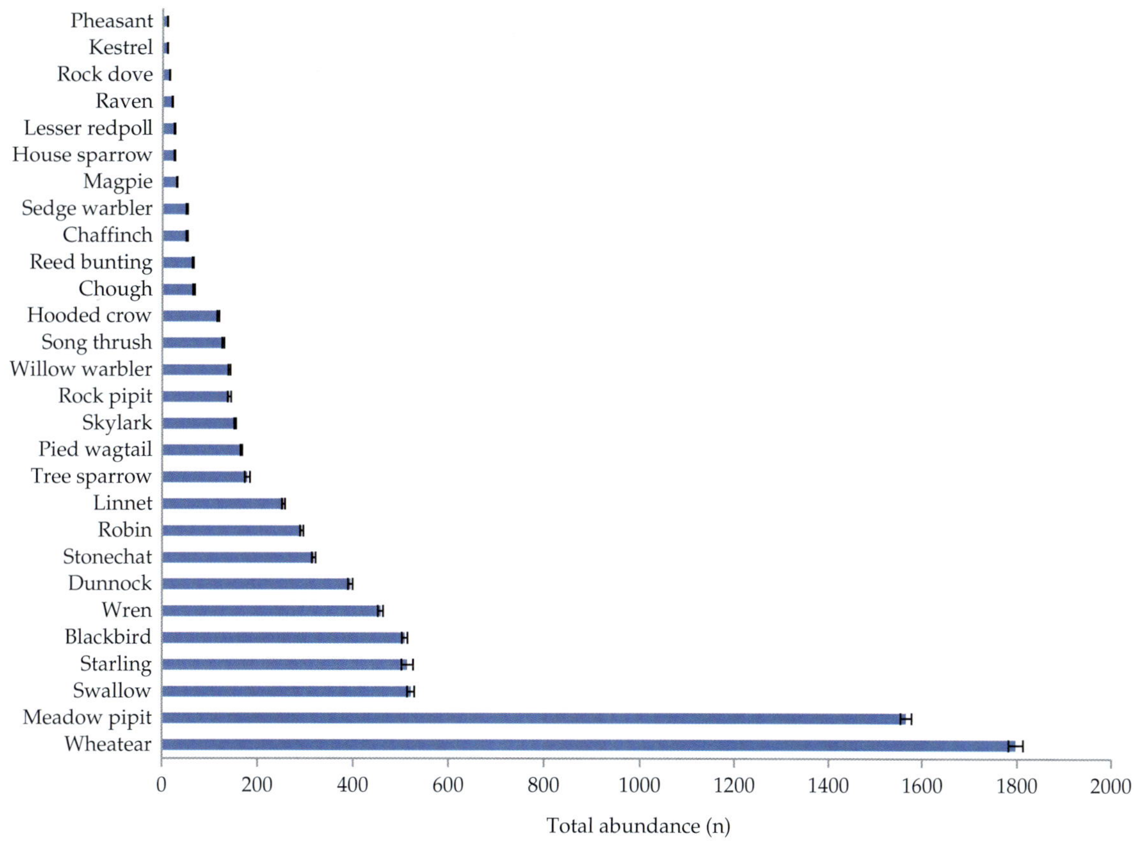

Fig. 8 Estimated number (+ SE) of individuals of each breeding land bird species on Clare Island in 2013.

relatively low total biomass (Fig. 9; see also Table 1). The wren and robin, despite being present in much larger numbers than most species, are, because of their low body weights, not within the top 50% of the heaviest species. Of the migrant species, only the northern wheatear is within the upper quartile, although the barn swallow is within the top 50% of the island's avian biomass.

Discussion

This study is, to our knowledge, the first to attempt to obtain a detailed estimate of the numerical abundance of land bird species on an Irish offshore island. Previously, Sharrock (1973) and Lysaght (2002) had provided quantitative data on the birds of Cape Clear (Cork) and the Aran Islands (Galway), respectively. Nevertheless, while more detailed, and based on an analysis using the program Distance, the Clare Island survey of June 2013 suffered from some limitations.

First, there was only one survey whereas ideally there should have been two: both an early visit (i.e. in April–May) and a later visit in June–July. However, for reasons explained in the Methods, an early visit in 2013 may not have been particularly informative as the spring and early summer of that year were both exceptionally cold and wet (Anonymous 2013a, 2013b, 2013c). The breeding season generally was delayed, and this fact was confirmed by the 16 to 18 June survey of Clare Island where very few recently fledged birds were detected but many adults were seen to be carrying food. It is argued therefore that the mid-June survey of 2013 did capture most of the nesting activity that year because of an exceptionally inclement spring and early summer. Of course, the impact(s) of the early suboptimal conditions on the normal demography of the island's breeding land birds could not be quantified as no baseline exists against which comparisons could be made.

A second limitation is that the transects were of different lengths. Retrospectively, transects of a predefined length would provide a standardised surveying unit that could potentially enable more accurate mean species densities to be calculated. With the data recorded in this study, it was not possible to perform statistically based comparisons on the densities of birds in different habitats.

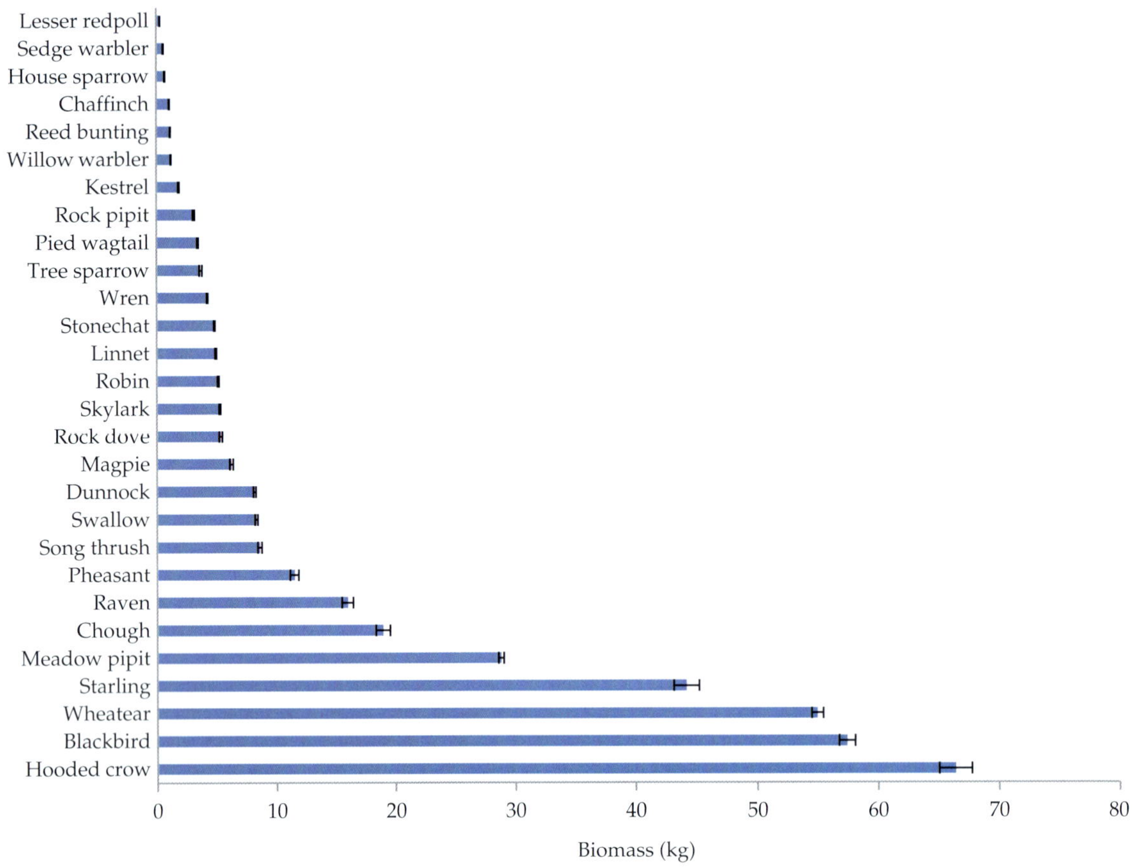

Fig. 9 Total biomass of individual species on Clare Island (± SE).

However, it is felt that with a different experimental design to the one followed here it should be possible to accomplish this in some areas of the island in the future.

The positions of the transects are a third difficulty. The terrain of Clare Island does not permit the implementation of multiple randomly placed transects of equal length throughout the entire island. The selection of the 23 'transect lines' was influenced by the presence of houses, walls, hedges and access to fenced-off, privately owned land. Therefore, some transects did not extend to all extremities of the island. In addition, the area around Lough Avullin was not surveyed in detail because this habitat and some of the surrounding land is difficult to traverse and is perhaps—like some of the woodland on the island—more suitably surveyed by the Point Count method.

Nevertheless, the transects covered a large survey area (19.7% of the entire island) and one that was representative of most of the component terrestrial habitats and, therefore, this study provides, for the first time, a baseline of numerical abundance against which future surveys can be compared. Furthermore, the implementation of a minimum of three distinct defined distance bands would provide for more robust estimates of density. These findings are discussed in more detail below.

Species richness

The 2013 breeding land bird survey detected 28 species as shown in Table 1 (approximately 20% of the total number of breeding bird species in Ireland) (Kelly 2008; Smiddy 2010; Kelly *et al.* 2014; Smiddy 2014). However, previous expeditions including those before the breeding season commenced, or as it was coming to a close, had detected additional species including the woodpigeon *Columba palumbus* Linnaeus, jackdaw *Corvus monedula* Linnaeus, grey wagtail *Motacilla cinerea* Tunstall, common cuckoo *Cuculus canorus* Linnaeus, sparrowhawk *Accipiter nisus* (Linnaeus), merlin *Falco columbarius* Linnaeus, twite *Carduelis flavirostris* (Linnaeus), collared dove *Streptopelia decaocto* (Frivaldszky) and sand martin *Riparia*

riparia (Linnaeus) (Barlee and Ruttledge 1945; Lloyd 1984; Gibbons *et al.* 1993; Forsyth and Allen 1997; Coombes 2003; Chapter 2). None of these species was detected in June 2013 but it is believed that some at least breed periodically on the island (see Chapter 2); it is also possible that some merely winter on the island (e.g. Lack 1986) or are passage migrants whose visits may overlap with the breeding season. Likewise, though not *sensu strictu* land birds, the moorhen *Gallinula chloropus* (Linnaeus) is also likely to be an irregular breeding species (Chapters 2 and 5). Therefore, the breeding land bird avifauna of Clare Island—as on all offshore islands—is in a state of flux. Some species have become extinct since the first survey (i.e. the corn bunting *Emberiza calandra* (Linnaeus), yellowhammer *Emberiza citrinella* Linnaeus and common whitethroat *Sylvia communis* Latham), while the twite, sand martin and sparrowhawk have departed the island more recently (Ussher 1912; Lloyd 1984; Forsyth and Allen 1997; Chapter 2). Relatively recent arrivals include the lesser redpoll, chaffinch, magpie *Pica pica* (Linnaeus), willow warbler (Lloyd 1984; Forsyth and Allen 1997; R. Cussen pers. comm.; this study) and goldcrest (Winters 2006). It is clear from the present study that many of the recently arrived species are now widely distributed and well established. Interestingly, both the lesser redpoll and willow warbler have increased dramatically in Ireland in recent years (Balmer *et al.* 2013), and while the latter appears to have become progressively more abundant, the common chiffchaff *Phylloscopus collybita* (Vieillot) does not breed (in fact, there are only two confirmed autumn records of the species on Clare Island; see Chapter 2).

As already mentioned, Clare Island has a relatively diverse land bird avifauna that is mostly composed of 'open' habitat-species and generalists. Woodland specialists are largely absent even though the area covered by maturing trees—both deciduous and conifer—is increasing. There is not much evidence of bird feeders or nest boxes on Clare Island and this may also explain, in part, why the Paridae are completely absent (though with a few non-breeding autumn records, Eoin McGreal pers.comm; see also Chapter 2). However, there is an emerging mosaic of garden habitats and most land bird species are associated with these. Human habitations and associated buildings such as outhouses and farmyard buildings provide nest sites for species such as the barn swallow, common starling, robin, tree sparrow and pied wagtail *Motacilla alba* Linnaeus (e.g. Ferguson-Less *et al.* 2011) and domestic stock including cattle, and stored fodder, provide feeding opportunities (e.g. Fennessy 2001; Holland *et al.* 2006; Henderson *et al.* 2007). The changes in the species composition of birds on Clare Island are discussed in greater detail in Chapters 2, 3, 5 and 6.

Numerical abundance

While the number and identity of the species detected in each transect is believed to be very accurate, there is less certainty (given the caveats mentioned above) about the estimated numerical abundances (Fig. 8). Because this is the first quantitative study of island land bird avifauna in

Vegetable gardens are a rich habitat for land birds. Potato drills and stooks of oats (the latter photographed in the autumn of 2005) are good feeding sites for birds. Robins *Erithacus rubecula* (Linnaeus), wrens *Troglodytes troglodytes* (Linnaeus), dunnock, *Prunella modularis* (Linnaeus), blackbird *Turdus merula* Linnaeus and song thrush *Turdus philomelos* C.L.Brehm, all use these habitats, and grain eaters such as the tree *Passer montanus* (Linnaeus) and house sparrows *Passer domesticus* (Linnaeus) will also be seen. Wheatears *Oenanthe oenanthe* (Linnaeus), chaffinches *Fringilla coelebs* Linnaeus, starlings *Sturnus vulgaris* Linnaeus, and the crows—magpie *Pica pica* (Linnaeus) and, hooded crow *Corvus cornix* Linnaeus—will also be present and swallows *Hirunda rustica* Linnaeus will criss-cross the vegetable gardens in search of flying insects. Photo: T. Kelly

Ireland, it is not possible to compare our results with those of previous studies. In addition, many of the recent studies of the avifaunas of different habitats in Ireland have focused primarily on woodland (e.g. Wilson *et al.* 2006, 2012); a community which, as mentioned, is largely absent from Clare island.

The density of the land bird avifauna is overwhelmingly dominated by open habitat species (meadow pipit and northern wheatear), which in turn reflects the availability of mostly unstructured blanket bog and moorland on the island, some of which is traversed by dry-stone walls. However, there are also areas of grassland, some of which is semi-natural (see Ryle 2013). Interestingly, given the abundance of the meadow pipit, particularly 'on moorland and heathland' (Glue and Morgan 1972; Brooke and Davies 1987), it is striking that the common cuckoo appears to have become extinct even though there are a few isolated records (Chapter 2).

In terms of diet, ground feeding invertebrate predators (e.g. meadow pipit, northern wheatear, dunnock, wren, blackbird, robin, stonechat *Saxicola torquatus* (Linnaeus), hooded crow and song thrush, as well as aerial insectivores (barn swallow)) dominate avian density.

As explained, the land bird avifauna of Clare Island is composed mainly of ground feeding invertebrate predators. Thus, earthworms, adult and immature insects, spiders and gastropod molluscs appear to form the mainstay of the land bird diet on Clare Island during the breeding season. It is unknown what proportion of this prey base are primary or secondary consumers (except for, of course, the spiders). Unfortunately, it has not been possible to research the ecological interactions and probable dietary overlap between these potentially competing passerine species.

It is remarkable how few herbivorous bird species are present on Clare Island and numerically they constitute a relatively small fraction of the overall island total (Fig. 8). The rock dove, linnet and lesser redpoll can be considered primarily herbivorous and feed their chicks with plant material either directly (linnet and lesser redpoll) or via a protein-rich 'milk' (rock dove) (Newton 1967; Moorcroft *et al.* 2006; Holland *et al.* 2006; BWP*i* V2). Moreover, while the linnet appears to provision its nestlings exclusively with seeds (Moorcroft *et al.* 2006), the lesser redpoll diet appears to be more mixed with, at some locations, variable amounts of insects being also fed to chicks (BWP*i* V2).

Apart from the corvid species, the pheasant appears to be the only definite omnivore, but its chicks, like the waders and wildfowl, are precocial and can provision themselves, at least in part, from when they hatch. However, they initially feed on arthropods and do not feed on seeds until six weeks old (BWP*i* V2).

Vertebrate predators include the peregrine falcon *Falco peregrinus* Tunstall, common kestrel *Falco tinnunculus* Linnaeus and sparrowhawk, which bred in most years between 1989 and 1999 (R.E. Cussen pers. comm.), but not since (see Chapter 2). Kestrels are regularly seen and have been proved to breed in most years since 2013. Less certain is the merlin, which has been recorded several times in recent years, but no nest has ever been found.

Biomass

The estimate of the biomass of the land birds on Clare Island (Fig. 9) is obviously crucially based on the accuracy of the population estimates (Fig. 8) (Dolton and Brooke 1999). We estimate the total land bird biomass on Clare Island to be >376.7kg, which amounts to 23.6kg per km^2. There appears to be very few, if any, comparable estimates and none that we are aware of that relate to islands in the North Atlantic. However, Nuorteva (1971), working in southern Finland, quotes a biomass of 21.9 and 22.9kg/km^2 in forested and agricultural lands respectively. These latter figures are broadly in line with, but lower than the biomass calculated in the current study. It is also worth noting that these population density estimates were obtained using different methodologies than those employed in the current study, so caution is needed when making direct comparisons. In fact, a variety of different methods have been used to estimate total biomass, the existence energy requirement of that biomass (i.e. the daily energy demands of each bird species), and how it interacts with the standing crop of available 'energy' (e.g. Karr 1968; Appelbaum and Haney 1981; Southwood *et al.* 1986; Hutto 1990; Evans *et al.* 2006; Vorisek *et al.* 2010).

The biomass of avian herbivores is approximately 10.7kg or 2.8% of the island's total. It would therefore appear that the Net Primary Production (NPP) of the island is mostly consumed by domestic animals—cattle and sheep, the rabbit *Oryctolagus cuniculus* (Linnaeus) the Irish hare *Lepus timidus hibernicus* (Bell) and small mammals such as rodents. It is of interest, however, that flocks of barnacle geese *Branta leucopsis* (Bechstein) (n='70–100'), an obligate herbivore (e.g. Prop and Black 1998), used to winter on Clare Island (e.g. Ruttledge 1950) and apparently did so up to the early 1980s (Lack 1986), but not recently (Balmer *et al.* 2013)—though see Chapter 2. A small number of Canada geese *Branta canadensis* (Linnaeus) were introduced, probably in 2014.

The non-herbivorous birds have a biomass of 366kg, and these must depend on the other primary and secondary consumers, that is, arthropods, molluscs and other invertebrates such as the Oligochaeta. In addition, the island's extensive low tide coastal habitats will provide prey for both the hooded crow and raven (e.g. Berrow *et al.* 1992; Smiddy 2016). It would be of interest to model the energy flow to the island's land birds in what must be a relatively simple system.

Avian land bird community on Clare Island

The breeding land bird community at Clare Island can be compared with others in Ireland where the numerical abundance of the component species has been estimated using modern techniques (e.g. Lysaght 1989; Bracken and Bolger 2006; Wilson *et al.* 2006, 2012; Bracken *et al.* 2008; Sweeney *et al.* 2010, 2012; Bracken and Smiddy 2012; Cummins *et al.* 2012; Nairn and O'Halloran 2012; Graham *et al.* 2015) and these data are therefore broadly comparable with the present study. However, while it should be emphasised that no comparable offshore island data were available, there are noticeable differences between the land bird community on Clare Island and those occupying what might be considered broadly similar habitats that have been surveyed recently elsewhere on the island of Ireland (e.g. Bracken *et al.* (2008), Bracken and Smiddy (2012) and Graham *et al.* (2015)).

In their detailed transect-based study, Bracken *et al.* (2008) showed that the meadow pipit and skylark were, together, the dominant species in Irish peatland habitats accounting for 'almost three quarters of all birds recorded during this study', and 'in raised and blanket bog habitats these figures rose to over 90% in each case'. However, on Clare Island, where there is a mosaic of relatively unstructured ground layer vegetation (Ryle 2013) (including farmland), the combined calculated fraction of meadow pipit and skylark was only 21.45% of the estimated >8,000 total number of land birds.

The effects of intense grazing is to remove 'structure' in the vegetation. The impact on bird species diversity is obvious. However, wheatears, meadow pipits *Anthus pratensis* (Linnaeus) feed in these areas as do flocks of choughs *Pyrrhocorax pyrrhocorax* (Linnaeus), starlings, and the occasional raven *Corvus corax* Linnaeus and hooded crow. Swallows will be seen flying low over the ground in their perpetual search for unwary flying insects. In the autumn it is possible that these very large areas of open closely cropped grassland would attract rare North American shorebirds such as the buff-breasted sandpiper *Tryngitis subruficollis* (Vieillot) and American golden plover *Pluvialis dominica* (P.L.S Müller). In the late autumn, winter and spring it is likely that the merlin *Falco columbarius* Linnaeus, will hunt these areas for meadow pipits and perhaps skylarks as well as starlings. Photos: T. Kelly

Moreover, while meadow pipit numbers generally exceed those of skylarks on bog and moorland habitats by 2 to 3: 1 in Ireland (Bracken *et al.* 2008; Bracken and Smiddy 2012), on Clare Island the former outnumbered the latter by at least 10: 1—a figure approaching the 18: 1 ratio shown to occur on fens (Bracken and Smiddy 2012). Similarly, Graham *et al.* (2015), using the point count technique, estimated meadow pipit to skylark ratios of 10: 1 in unforested wet grassland and 2.7: 1 in peatland.

It is notable however that in this latter study northern wheatear numbers were 'too low for inclusion in statistical analysis' and the species does not figure in the lists of the more abundant bird species found on bogs and fens given by Bracken *et al.* (2008) or in the comprehensive review of Bracken and Smiddy (2012), possibly in part because of a paucity of suitable breeding sites.

In marked contrast, therefore, the northern wheatear was found to be the most abundant species during the transect-based census on Clare Island, a fact that highlights the major difference between this land bird community and those described by Bracken *et al.* (2008), Bracken and Smiddy (2012) and Graham *et al.* (2015). Even allowing for possible errors in the census (e.g. inadvertent inclusion of recently fledged birds, and the possibly unjustified extrapolation from the numbers found in the transect survey areas to the island as a whole) and the known difficulties associated with estimating the number of breeding adults of this species (Sellers 2006; Gillings and Newson 2007; Fulton 2010; Balmer *et al.* 2013), it would appear that the numbers of northern wheatear—an amber-listed bird species of conservation concern (BOCC; Colhoun and Cummins 2013)—are exceptionally high on Clare Island.

In marked contrast to the meadow pipit that appears to be resident or a partial migrant in Ireland (BWP*I* V2; Wernham *et al.* 2002), the northern wheatear is a long distance trans-Saharan migrant. The wheatear is mostly a single brooded species (BWP*i* V2) with a clutch size of five eggs whereas the meadow pipit produces two broods each with four eggs (e.g. Ferguson-Lees *et al.* 2011). The meadow pipit is probably resident and sedentary in the west of Ireland, remaining there over the winter months (Lack 1986; Balmer *et al.* 2013), whereas the wheatear vacates its breeding territories from August through to April (Wernham *et al.* 2002). Both species have shown marked declines in the size of their breeding populations as well as contractions in their geographical distribution (Balmer *et al.* 2013). Nevertheless, the meadow pipit maintains a population of >98 birds per km^2 on Clare Island and the wheatear density is even higher at >112 per km^2—an exceptional number (but not the highest) compared to the data given by Gillings and Newson (2007).

Research in Sweden has shown that northern wheatear nests are depredated by ground predators such as mustelids—including the badger *Meles meles* (Linnaeus), and other carnivores such as red fox *Vulpes vulpes* (Linnaeus) as well as by snakes (e.g. the adder *Vipera berus* (Linnaeus)) (Arlt *et al.* 2008; Schneider *et al.* 2012)—all of which are absent from Clare Island. While domestic cats *Felis catus* Linnaeus, brown rats *Rattus norvegicus* (Berkenhout), hooded crows and ravens are all present on Clare Island, it appears that the pressure exerted there on wheatears and indeed other ground nesting bird species from predators may be considerably less than that experienced on the island of Ireland as a whole, though Robert Cussen (pers.comm) considers the domestic cat to be a major predator on Clare Island. Some predator populations are increasing in Ireland, most notably the pine marten *Martes martes* (Linnaeus) and American mink *Neovison vison* Schreber—the latter has reached some offshore islands but not, so far as is known, Clare Island (Lysaght and Marnell 2016; Colin Lawton in press). This may mean that offshore islands could become important refuges for some vulnerable ground, and near ground, nesting bird species in Ireland.

These observations suggest that a controlled long-term study of the breeding biology and demography of the northern wheatear on the mainland and Clare Island might be worthwhile. Such a study should also undertake habitat specific censuses, assess the impact of predators and conduct detailed investigations of the diet of the species during different phases of the breeding season (see especially Arlt *et al.* 2008; Schneider *et al.* 2012; Seward *et al.* 2013). Finally, it would be interesting to discover (through satellite or similar tags) what route(s) the Clare Island wheatears take to and from their wintering quarters in Africa.

Summary and conclusions

A total of 28 land bird species were detected and censused in June 2013. Numerically speaking the island's avifauna is dominated by open habitat species such as the meadow pipit and northern wheatear, with the latter being exceptionally numerous. The very high density of ground nesting passerine species suggests that the relatively low numbers of ground predators on Clare Island may mean that it acts as a refuge (as do many other islands), where survival and reproductive success are higher than on the mainland of Ireland, though further research is necessary to confirm this. Most of the other species are habitat generalists that are often closely associated with man-made features such as hedgerows, walls, outhouses and gardens. Herbivorous birds form only a very small fraction (>3%) of the island's terrestrial avian biomass. Therefore, the populations of the breeding species must be dependent on invertebrate prey with which to provision their nestlings. This relatively simple food web would be interesting to model in parallel with a detailed study of the demography of the commoner species.

Acknowledgements

We are indebted to the referee for the detailed critique and constructive comments; to Pat Smiddy for his masterful edits; Ms R. Hutchinson-Kelly for her help with the fieldwork and to both her and Stuart Warner for their assistance with the production of this paper. This study was privately funded.

REFERENCES

Alldredge, M.W., Simons, T.R., and Pollack, K.H. 2007 Factors affecting aural detections of songbirds. *Ecological Applications* **17**, 948–55.

Anonymous 2013a *Monthly Weather Bulletin, April 2013*. No. 324. Glasnevin. Met Éireann.

Anonymous 2013b *Monthly Weather Bulletin, May 2013*. No. 325. Glasnevin. Met Éireann.

Anonymous 2013c *Monthly Weather Bulletin, June 2013*. No. 326. Glasnevin. Met Éireann.

Appelbaum, S. and Haney, A. 1981 Bird populations before and after wildfire in a Great Lakes pine forest. *Condor* **83**, 347–54.

Arlt, D., Forslund, P., Jeppsson, T. and Part, T. 2008 Habitat-specific population growth of a farmland bird. *PLoS One* **3**, e3006.

Balmer, D., Gillings, S., Caffrey, B., Swann, B., Downie, I. and Fuller, R. 2013 *Bird Atlas 2007–2011: the breeding and wintering birds of Britain and Ireland*. Thetford. BTO Books.

Barlee, J. and Ruttledge, R.F. 1945 Notes on the present status of birds on Clare Island. *Irish Naturalists' Journal* **8**, 311–3.

Berrow, S., Kelly, T.C. and Myers, A.A. 1992 The diet of coastal breeding hooded crows *Corvus corone cornix*. *Ecography* **15**, 337–46.

Berry, R.J. 2009 *The Natural History of Islands*. New Naturalist. London. Collins.

Bibby, C.J., Burgess, N.D. and Hill, D.A. 1992 *Bird Census Techniques*. London. Academic Press.

Bibby, C.J., Burgess, N.D., Hill, D.A. and Mustoe, S. 2000 *Bird Census Techniques*. Second edition. London. Academic Press.

BWP *Birds of the Western Palaearctic Interactive*. BWPi Version 2.

Blondel, J. 1985 Breeding strategies of the blue tit and coal tit (*Parus*) in mainland and island Mediterranean habitats: a comparison. *Journal of Animal Ecology* **54**, 531–56.

Blondel, J. 2000 Evolution and ecology of birds on Islands: trends and prospects. *Vie et Milieu* **50**, 205–20.

Blondel, J., Perret, P., Maistre, M. and Dias, P.C. 1992 Do harlequin Mediterranean environments function as source or sink for blue tits (*Parus caeruleus* L.). *Landscape Ecology* **6**, 213–9.

Bracken, F. and Bolger, T. 2006 Effects of set-aside management on birds breeding in lowland Ireland. *Agriculture, Ecosystems and Environment* **117**, 178–84.

Bracken, F., McMahon, B.J. and Whelan, J. 2008 Breeding bird populations of Irish peatlands. *Bird Study* **55**, 169–78.

Bracken, F. and Smiddy, P. 2012 Lowland bogs, Fens and Reed swamps. In Nairn, R. and O'Halloran, J. (eds) *Bird Habitats in Ireland*. 73–89. Cork. Collins Press.

Brooke, M.de L. and Davies, N.B. 1987 Recent changes in host usage by cuckoos *Cuculus canorus* in Britain. *Journal of Animal Ecology* **56**, 873–83.

Buckland, S.J., Anderson, D.R., Burnham, K.P., Laake, J-L., Bochers, D.L. and Thomas, L. 2001 *Introduction to Distance Sampling: estimating abundance of biological populations*. Oxford. Oxford University Press.

Burnham, K.P. and Anderson, D.R. 1984 The need for distance data in transect counts. *Journal of Wildlife Management* **48**, 1248–54.

Colhoun, K. and Cummins, S. 2013 Birds of Conservation Concern in Ireland 2014–2019. *Irish Birds* **9**, 523–44.

Coombes, R.H. 2003 *Clare Island, Co. Mayo. Birds noted on a 3 day visit – 29/4 to 1/5/2000*. Unpublished report.

Cummins, S., O'Connell, P. and Mee, A. 2012 Upland heath and blanket bog. In Nairn, R. and O'Halloran, J. (eds) *Bird Habitats in Ireland*. 90–104. Cork. Collins Press.

Dolton, C.S. and Brooke, M.de L. 1999 Changes in the biomass of birds breeding in Great Britain 1968–88. *Bird Study* **46**, 274–8.

Dunning, J.B. (ed) 2007 *CRC Handbook of Avian Body Masses*. Second edition. Boca Raton. CRC Press.

Evans, K.L., James, N.A. and Gaston, K.J. 2006 Abundance, species richness and energy availability in the North American avifauna. *Global Ecology and Biogeography* **15**, 372–85.

Fennessy, G. 2001 Aspects of the ecology of the robin *Erithacus rubecula* L. Ph.D. thesis. Cork. National University of Ireland.

Ferguson-Lees, J., Castell, R. and Leech, D. 2011 *A Field Guide to Monitoring Nests*. Thetford. British Trust for Ornithology.

Forsyth, I. and Allen, P. 1997 *Report to the Royal Irish Academy on the Breeding Birds of Clare Island, Co. Mayo, 1996 and 1997*. 9 pages. Unpublished report to Royal Irish Academy.

Fulton, D. 2010 The breeding population of northern wheatears at Clee Hill, Shropshire, 1998–2009. *British Birds* **103**, 223–8.

Gibbons, D.W., Reid, J.B. and Chapman, R.A. (eds) 1993 *The New Atlas of Breeding Birds in Britain and Ireland: 1988–1991*. London. Poyser.

Gillings, S. and Newson, S.E. 2007 Breeding population estimate for northern wheatear in Britain. *British Birds* **100**, 179–84.

Glue, D. and Morgan, R. 1972 Cuckoo hosts in British habitats. *Bird Study* **19**, 187–92.

Gottschalk, T.K. and Huettmann, F. 2010 Comparison of distance sampling and territory mapping methods for birds in four different habitats. *Journal of Ornithology* **152**, 421–9.

Graham, C.T., Wilson, M.W., Gittings, T., Kelly, T.C., Irwin, S., Quinn, J.L. and O'Halloran, J. 2015 Implications of afforestation for bird communities: the importance of preceding land-use type. *Biodiversity Conservation* DOI 10.1007/s10531–015–0987–4.

Henderson, I., Holt, C. and Vickery, J. 2007 National and regional patterns of habitat association with foraging barn swallows *Hirundo rustica* in the UK. *Bird Study* **54**, 371–7.

Holland, J.M., Hutchison, M.A.S., South, B. and Aebischer, N.J. 2006 A review of invertebrate and seed bearing plants as food of farmland birds in Europe. *Annals of Applied Biology* **148**, 49–71.

Hutchinson, C.D. 1989 *Birds in Ireland*. Calton. Poyser.

Hutto, R.L. 1990 Measuring the availability of food resources. *Studies in Avian Biology* **13**, 20–8.

Jurasinski, G., Retzer, V. and Beirerkuhnlein, C. 2009 Inventory, differentiation and proportional diversity: a consistent terminology for quantifying species diversity. *Oecologia* **159**, 15–26.

Karr, J.R. 1968 Habitat and avian diversity on strip-mined land in east-central Illinois. *Condor* **70**, 348–57.

Kelleher, K.M. and O'Halloran, J. 2006 Breeding biology of the song thrush *Turdus philomelos* in an island population. *Bird Study* **53**, 142–55.

Kelly, T.C. 2008 The origin of the avifauna of Ireland. In Davenport, J.L., Sleeman, D.P. and Woodman, P.C. (eds) *Mind the Gap: Post Glacial colonisation of Ireland*. 97–107. Belfast. Irish Naturalists' Journal Special Supplement.

Kelly, T.C., Smiddy, P. and Graham, C.T. 2014 Origins of the Irish avifauna – a perspective from Island Biography Theory. In Sleeman, D.P., Carlsson, J. and Carlsson, J.E.L. (eds) *Mind the Gap II: new insights into the Irish Postglacial*. 44–52. Belfast. Irish Naturalists' Journal.

Lack, P.C. 1986 *The Atlas of Wintering Birds in Britain and Ireland*. Calton. Poyser.

Lloyd, C.S. 1984 The birds of Clare Island, Co. Mayo in June 1982. *Irish Naturalists' Journal*. **21**, 212–6.

Lysaght, L. 1989 Breeding bird populations of farmland in mid-west Ireland. *Bird Study* **36**, 91–8.

Lysaght, L. 2002 *An Atlas of Breeding Birds of the Burren and the Aran Islands*. Dublin. BirdWatch Ireland.

Lysaght, L. and Marnell, F. (eds) 2016 *Atlas of Mammals in Ireland 2010–2015*. Waterford. National Biodiversity Data Centre.

Moorcroft, D., Wilson, J.D. and Bradbury, R.B. 2006 Diet of nestling linnets *Carduelis cannabina* on lowland farmland before and after agricultural intensification. *Bird Study* **53**, 156–62.

Nairn, R. and O'Halloran, J. (eds) 2012 *Bird Habitats in Ireland*. Cork. Collins Press.

Newton, I. 1967 The adaptive radiation and feeding ecology of some British finches. *Ibis* **109**, 33–98.

Nuorteva, P. 1971 The synanthropy of birds as an expression of the ecological cycle disorder caused by urbanisation. *Annales Zoologici Fennici* **8**, 547–53.

O'Connell, S., Irwin, S., Wilson, M.W., Sweeney, O., Kelly, T.C. and O'Halloran, J. 2012 How can forest management benefit bird communities? Evidence from eight years of research in Ireland. *Irish Forestry* **69**, 44–57.

O'Halloran, J., Nairn, R., Smiddy, P. and Crowe, O. 2012 Bird habitats: a synthesis and future perspectives. In Nairn, R. and O'Halloran, J. (eds) *Bird Habitats in Ireland*. 247–68. Cork. Collins Press.

Parkin, D.T. and Knox, A.G. 2010 *The Status of Birds in Britain and Ireland*. London. Christopher Helm.

Prop, J. and Black, J.M. 1998 Food intake, body reserves and reproductive success of barnacle geese *Branta leucopsis* staging in different habitats. *Norsk Polarinstitutt Skrifter* **200**, 175–93.

Reed, T. 1980 Turnover frequency in island birds. *Journal of Biogeography* **7**, 329–35.

Reed, T. 1981 The number of breeding landbird species on British islands. *Journal of Animal Ecology* **50**, 613–24.

Ruttledge, R.F. 1950 A list of the birds of the counties of Galway and Mayo: showing their status and distribution. *Proceedings of the Royal Irish Academy* **52**B, 315–81.

Ryle, T. 2013 Vegetation-Environment interactions on Clare Island. In D. Synnott (ed.) *New Survey of Clare Island*. Volume **7**, 27–178. Dublin. Royal Irish Academy.

Schneider, N.A., Low, M., Arlt., D. and Part, T. 2012 Contrast in edge vegetation structure modifies the predation risk of natural ground nests in an agricultural landscape. *PloS One* **7**, e31517.doi 10.1371/journal.pone.00131517.

Sellers, R.M. 2006 Breeding population estimate for northern wheatears in Britain. *British Birds* **99**, 533–35.

Seward, A.M., Beale, C.M., Gilbert, L., Hefin Jones, T. and Thomas, R.J. 2013 The impact of increased food availability on survival of a long-distance migratory bird. *Ecology* **94**, 221–30.

Sharrock, J.T.R. 1973 *The Natural History of Cape Clear Island*. Berkhamsted. Poyser.

Smiddy, P. 2010 A systematic list of Irish Birds. *Irish Birds* **9**, 130–4.

Smiddy, P. 2014 Breeding birds in Ireland: success and failure among colonists. In Sleeman, D.P., Carlsson, J. and Carlsson, J.E.L. (eds) *Mind the Gap II: new insights into the Irish Postglacial*. 89–99. Belfast. Irish Naturalists' Journal.

Smiddy, P. 2016 Diet of coastal breeding ravens *Corvus corax* in east County Cork. *Irish Birds* **10**, 335–8.

Smiddy, P., O'Halloran, J., O'Mahony, B. and Taylor, A.J. 1995 The breeding biology of the dipper *Cinclus cinclus* in southwest Ireland. *Bird Study* **42**, 76–81.

Southwood, T.R.E., Brown, V.K., Reader, P.M. and Green, E.E. 1986 The use of different stages of succession by birds. *Bird Study* **33**, 159–63.

Sweeney, O.F., Kelly, T.C., Irwin, S., Wilson, M.W. and O'Halloran, J. 2012 Woodlands, forest and scrub. In Nairn, R. and O'Halloran, J. (eds) *Bird Habitats in Ireland*. 105–23. Cork. Collins Press.

Sweeney, O.F., Wilson, M.W., Irwin, S., Kelly, T.C. and O'Halloran, J. 2010 Breeding bird communities of second-rotation plantations at different stages of the forest cycle. *Bird Study* **57**, 301–14.

Thomas, L., Buckland, S.T., Burnham, K.P., Anderson, D.R., Laake, J.L., Borchers, D.L. and Strindberg, S. 2013 Distance sampling. *Encyclopedia of Environmetrics*. Second edition. Doi:10.1002/97 80470057339.vad033. pub2. Accessed 30 March 2017.

Thomas, L., Buckland, S.T., Rexstad, E.A., Laake, J.L., Strindberg, S., Hedley, S.L., Bishop, J.R.B., Marques, T.A. and Burnham, K.P. 2010 Distance software: design and analysis of distance sampling surveys for estimating population size. *Journal of Applied Ecology* **47**, 5–14.

Ussher, R.J. 1912 A biological survey of Clare Island in the county of Mayo, Ireland and adjoining district. *Proceedings of the Royal Irish Academy* **31**, Sec.11, Part 20.

Vorisek, P., Jiguet, F., van Strien, A., Skorpilova, J., Kvalanova, A. and Gregory, R.D. 2010 Trends in abundance and biomass of widespread European farmland birds: how much have we lost? *European trends in farmland birds. BOU Proceedings Lowland Farmland Birds III.* http//www.bou.org.UK/bouproc-net/yb3/vorisek et al. pdf.

Wernham, C., Toms, M., Marchant, J., Clark, J., Siriwardina, G. and Baillie, S. (eds) 2002 *The Migration Atlas: movement of birds of Britain and Ireland*. London. Poyser.

Wilson, M.W., Gittings, T., Pithon, J., Kelly, T.C., Irwin, S. and O'Halloran, J. 2012 Bird diversity of afforestation habitats in Ireland: current trends and likely impacts. *Biology and Environment: Proceedings of the Royal Irish Academy* **112**B, 55–68.

Wilson, M.W., Pithon, J., Gittings, T., Kelly, T.C., Giller, P.S. and O'Halloran, J. 2006 Effects of growth stage and tree species composition on breeding bird assemblages of plantation forests. *Bird Study* **53**, 225–36.

Winters, P. 2006 A summer bird survey of Clare Island. M.Sc. thesis. Dublin. Centre for the Environment, Trinity College.

CHAPTER 5

WATERBIRDS – DUCKS, RAILS AND WADERS ON CLARE ISLAND

T.C. Kelly, U. Keating, P. Winters, R.H. Coombes, S. Warner, E. McGreal and P. Smiddy

ABSTRACT

This paper, based in part on a transect study carried out in June 2013, describes the community of wading birds, ducks and rails breeding on Clare Island (1,598 ha) and compares the findings with those of the first survey conducted in 1909–11 (Ussher 1912). Five wading bird species now breed on the island—the same number as was discovered during the first survey. Thus, the oystercatcher *Haematopus ostralegus* Linnaeus, ringed plover *Charadrius hiaticula* Linnaeus, northern lapwing *Vanellus vanellus* (Linnaeus), common snipe *Gallinago gallinago* (Linnaeus) and common sandpiper *Actitis hypoleucos* (Linnaeus) are all present on the island today as they were over 100 years ago. This number of wading bird taxa represents 50% of the current species richness of Charadriidae and Scolapacidae that regularly breed in Ireland. Currently, none of the species is numerically abundant and this also seems to have been the situation during the first survey, despite the absence of most of the mammalian predators known to occur in Ireland. However, waterbodies are surprisingly few on Clare Island and the largest—Lough Avullin—has changed from an open water lake habitat of some 1.5ha in 1909, to an impenetrable reed swamp, over the 100+ year interval since the first survey. This may explain why Clare Island holds only one duck species—the mallard *Anas platyrhynchos* Linnaeus—and only one rail—the moorhen *Gallinula chloropus* (Linnaeus)—though the latter only breeds there sporadically. The water rail *Rallus aquaticus* Linnaeus, which appears to have been relatively common during the time of the first expeditions, has not been recorded since, though the large and dense reed bed at Lough Avullin looks an ideal breeding habitat for this species, but one which is exceptionally difficult to survey. Overall, the low diversity of water birds on Clare Island shows that area *per se* is a poor predictor of species richness in this group.

Introduction

Aquatic birds associated with insular lentic and lotic freshwater habitats do not appear to be good candidates for an Island Biogeography Theory-type analysis. Thus, the species richness of *inter alia*, ducks (Anatidae), rails (Rallidae) and waders (Charadriidae and Scolopacidae) will be dependent on both the amount and variety of freshwater habitats on an island. However, both these variables may vary independently of the area of an island. For example, some relatively large archipelagos such as the Blasket Islands (Kerry) have little or no freshwater habitats whereas much smaller groups such as the Inishkeas (Mayo) have an open body of water which provides a breeding habitat for 'freshwater-requiring' species (Cabot 1963).

Lochán at Maum where ringed plover *Charadrius hiaticula* Linnaeus bred in 2013. Lapwing *Vanellus vanellus* (Linnaeus) were seen here during the late breeding season in previous years. The overall habitat is ideal for wheatears. In an autumn with strong westerly winds it is likely that a North American shorebird such as a buff breasted sandpiper *Tryngitis subruficollis* (Vieillot) would be found at this site. Photo: T. Kelly

While most aquatic bird groups are exposed to many contemporary threats including drainage, intensification in agriculture, water extraction, hunting, recreational disturbance, pollution and climate change, the waders—plovers and sandpipers—appear, in particular, to be experiencing major population declines on both sides of the North Atlantic (e.g. Heath *et al.* 2000; Thorup 2006; Macdonald and Bolton 2008; Brown *et al.* 2015; Kirwan *et al.* 2015; Colhoun *et al.* 2015; Munro 2017). Wading birds, however, have exceptional life histories in that they exhibit deferred maturity, low clutch sizes but relatively high adult survival (Sandercock 2003). It is argued, therefore, that population declines are mainly caused by sustained breeding failure rather than increased adult mortality (e.g. Macdonald and Bolton 2008). The role of predators of both eggs and chicks has been highlighted as a major cause of breeding failure (Macdonald and Bolton 2008) though other factors may also be involved (e.g. Calladine *et al.* 2014).

Island populations of waterfowl and especially wading birds may be particularly interesting because of the absence, or scarcity, of mammalian ground predators. The remarkably destructive impact on the exceptionally large and important breeding wader populations on the Outer Hebridean island of South Uist that followed the introduction of the European hedgehog *Erinaceus europaeus* Linnaeus in 1974 is now well known (Jackson and Green 2000; Jackson 2001; Jackson *et al.* 2004) and serves as a spectacular example of ground predator-inflicted losses on breeding charadriid and scolapacid populations (but see also Calladine *et al.* 2014).

These findings are relevant to the breeding waterfowl avifauna of Clare Island that was first described by Ussher (1912) and included five species of nesting wading birds, namely: the oystercatcher *Haematopus ostrelagus* Linnaeus, ringed plover *Charadrius hiaticula* Linnaeus, northern lapwing *Vanellus vanellus* (Linnaeus), common snipe *Gallinago gallinago* (Linnaeus) and common sandpiper *Actitis hypoleucos* (Linnaeus); one breeding duck species, the mallard *Anas platyrhynchos* Linnaeus; and two rails, the water rail *Rallus aquaticus* Linnaeus and moorhen *Gallinula chloropus* (Linnaeus). The dipper *Cinclus cinclus* (Linnaeus) subspecies *hibernicus* Hartert, a lotic habitat specialist, was recorded as a breeding species by Ussher (1912) and although there have been occasional subsequent sightings, intensive searches for nests have failed to locate any on Clare Island since the time of the first survey (see Chapter 2).

The early surveys of Ussher (1912) did not provide much in the way of distributional data on the breeding waders or wildfowl on Clare Island or give, other than in broad outline, an indication of their numerical abundance. Subsequent expeditions did provide some numerical and distributional data, particularly the censuses undertaken by Ian Forsyth and Pamela Allen who surveyed the island on 8 May 1996 and again on 18 to 20 July in the same year. Forsyth and Allen (1997) also surveyed Clare Island on 14 to 16 April 1997. Presence or absence data was collected during the first breeding atlas (Sharrock 1976), the winter atlas (Lack 1986), and subsequent breeding atlases (Gibbons *et al.* 1993; Balmer *et al.* 2013). Important distributional data was provided by Lloyd (1984) and by R.H. Coombes who surveyed the island in late April and early May 2000 (Coombes 2003). A detailed survey of Clare Island birds was carried out by Patricia Winters from late May to the end of June 2006. Winters (2006) divided the area of the island into 1km squares. Each square was surveyed using two transects, with each transect being surveyed for approximately one hour during surveys (the exception to this was some of the smaller coastal squares, where time spent was reduced according to size). The detailed field work was confined to the mornings and evenings in order to avoid times of low breeding activity (the hours between 12.30 and 15.00 hrs were therefore avoided). Counts of birds and observations of breeding indicators were recorded during visits, with habitat types recorded simultaneously (Winters 2006).

Quantitative data—albeit preliminary—was also collected on the aquatic birds of Clare Island during the 2013 breeding season when 23 transects were undertaken in mid-June (see Chapter 4). This is only the second attempt at estimating the numerical abundance of waterfowl and waders on Clare Island but because of time constraints the total area of potential wader breeding habitats on the island could not be surveyed by the transect method.

The aims of this brief study therefore are (i) to summarise the findings of the June 2013 transect studies of waders and waterfowl and (ii) to integrate this with all the survey data collected during the previous censuses undertaken by those named above. Strong emphasis is placed on the data collected by Winters (2006), due to the detail of this survey. In addition, we also focus on the species richness aspect as the data relates to an interval of over 100 years extending from the early twentieth century (Ussher 1912) to 2013.

Lochán at Toormore where mallard *Anas platyrhynchos* Linnaeus and lapwing bred in 2013; other inhabitants include a small flock of farmyard ducks. It is likely that this site would attract a few teal *Anas crecca* Linnaeus in the winter months. Photo: T. Kelly

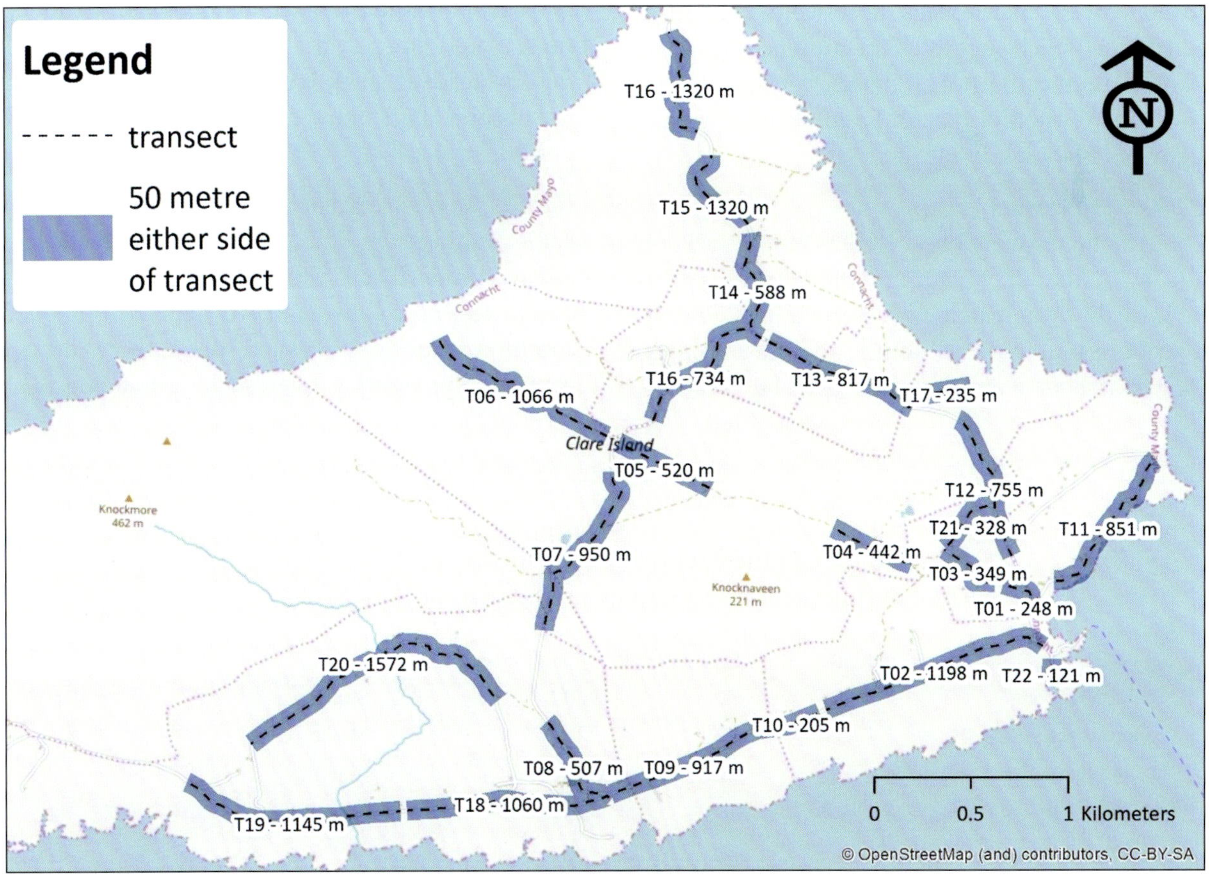

Fig. 1 Positions of 23 transects on Clare Island surveyed on 16 to 18 June 2013.

Materials and methods

Dates, times, and positions of 2013 transect surveys

All the fieldwork was conducted between 16 and 18 June 2013. Most transects were undertaken between 07.30 and 18.30 hrs, but one was commenced at 21.45 and completed at 22.15 hrs.

The positions of the 23 transects walked over the 16 to 18 June 2013 interval are shown in Fig. 1. The positions of the transects were selected with the constraint that most areas of fenced-off privately owned agricultural holdings were to be avoided; there are large areas of commonage on Clare Island and most of the fenced off areas can be readily surveyed from metalled roads or public boreens. The location of transects ensured that most, but not all, of the major wetland habitats on the island were surveyed.

Results

2013 survey results

It can be seen in Table 1 that a total of four wader and one duck species was found in the transect survey. An additional wader species—the common sandpiper—was discovered off transect near the mouth of the Owenmore River in Strake. Two adults issued alarm calls and were exhibiting behaviour, which suggested the presence of chicks, though these could not be detected. The approximate distribution of the 2013 aquatic bird species sightings is shown in Fig. 2.

2006 survey results

Three mallards were observed by Winters (2006), but evidence of breeding was not noted. A total of 82 oystercatchers were counted, widely distributed along the south and east coasts. Oystercatchers were determined to be breeding in five out of the fourteen squares in which they were observed. A total of 16 ringed plovers were observed, but although there was no evidence of breeding behaviours, suitable shingled nesting habitat was present in four of the six squares in which they were observed. Ten individuals were observed in one square which contains a stream with a large historic gravel floodplain, and which was observed to be a breeding ground for ringed plovers in the past (Winters 2006). Winters observed that it was likely that some of the ten individuals were breeding on this floodplain, or that there were more individuals unrecorded. Twelve northern lapwings were recorded during the survey. These occurred in one square on the east side of the island, with no signs of breeding observed. This square is one of the relatively few areas on the island where there is a considerable amount of marshy ground. Several small lakes with grassy islands were observed where lapwings were circling (Winters 2006). The survey recorded five common sandpipers, and again these were on the east side of the island. No evidence of breeding was observed in any of the three squares in which they were recorded, but the presence of a stony shoreline in each would suggest that breeding is possible, as this is an ideal chick habitat. Winters (2006) did not observe any common snipe during the survey period.

Results from surveys prior to 2006

Table 2 shows the data collected by Forsyth and Allen (1997). Their most detailed survey was conducted on 14 to 16 April 1997. Although a common snipe was detected, the authors did not consider the sighting referred to a breeding bird.

Table 1
Summary of the data relating to aquatic birds detected during the transect survey undertaken in June 2013.

Species	No. of individuals	No. of transects where present	Transect number where present	Areas where birds recorded
Northern lapwing	1	1	T.21	Strake
Oystercatcher	2	1	T.20	Strake
Ringed plover	3	1	T.18	Maum
Common snipe	3	3	T.12,13 & 14	Fawnglass, Maum, Ballytoohy, Beg
Mallard	2	1	T.21	Strake
Common sandpiper	2	0	Seen off transect	Strake

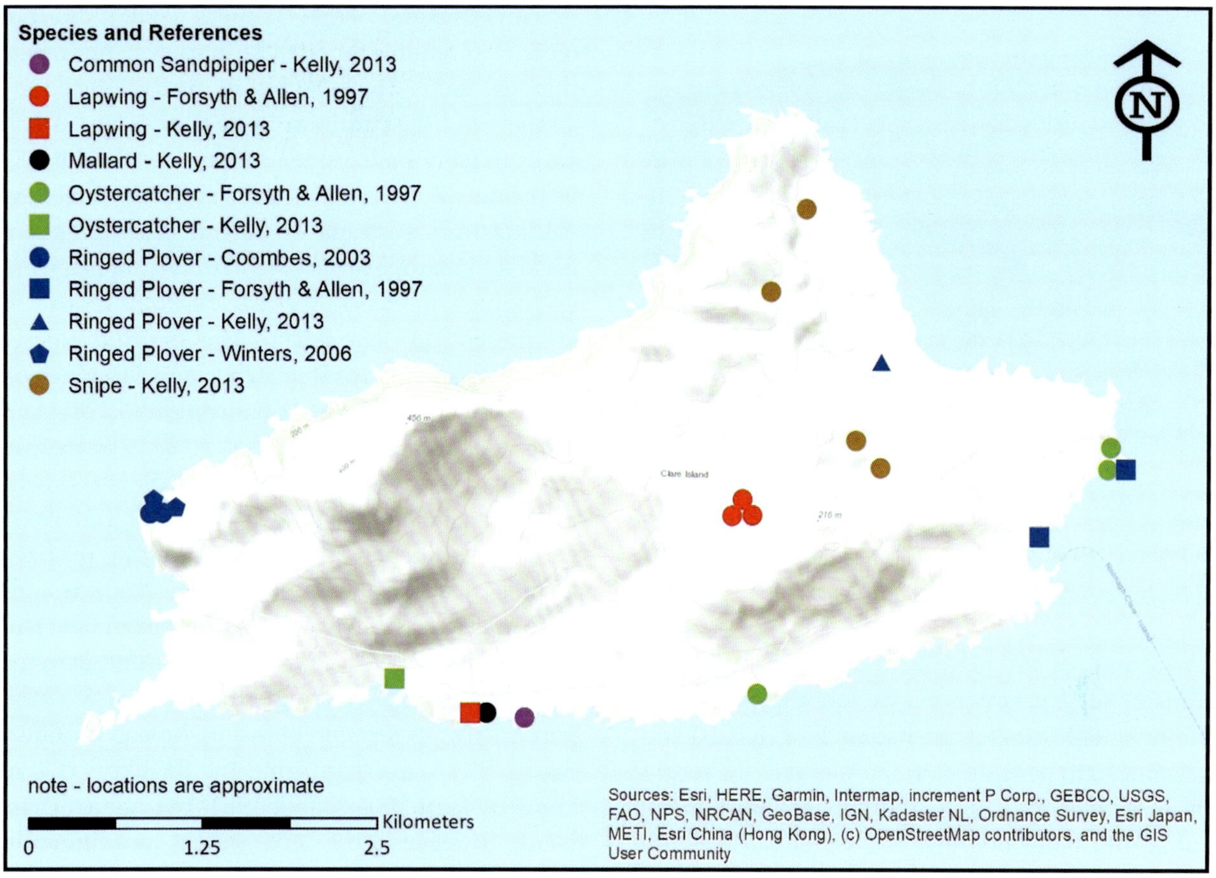

Fig. 2 Approximate positions of breeding wading bird and wildfowl species discovered during the surveys of Forsyth and Allen (1997), Coombes (2003), Winters (2006) and Kelly (2013; this study).

Table 2
Aquatic bird species detected during the Forsyth and Allen (1997) survey.

Species	Total number of pairs	Locations on Clare Island
Oystercatcher	2	Capnagower
Ringed plover	2	Capnagower
Lapwing	4	Capnagower & Lecarrow
Mallard	4	Lecarrow & Maum

Note: common snipe was seen on 16 April 1997 but was not believed to refer to a breeding record. Common sandpiper was not recorded in the 1997 survey.

No common sandpiper was detected during this survey (Forsyth and Allen 1997). However, estimates of the numbers of breeding oystercatchers (n=2 pairs), ringed plover (n=2 pairs) and lapwing ('at least 4 pairs') were obtained (Table 2). The distribution of the breeding waders and wildfowl discovered during the Forsyth and Allen (1997) survey is shown in Fig. 2.

The positions of the 2013 transects would suggest that most, if not all, of the locations of Forsyth and Allen's (1997) sightings were surveyed in 2013, though no registrations of any waders were made at or near Kinacorra Point in Capnagower in 2013; this location held both oystercatchers and ringed plover in 1997. Lloyd (1984) did not record common snipe or common sandpiper in her 1982 survey, but the locations of her mallard, lapwing and ringed plover sightings appear to have been in areas covered by the 23 transects conducted in 2013. Coombes (2003) did find common sandpiper but common snipe was not detected. However, his ringed plover and lapwing locations near the Signal Station were not surveyed during the 2013 transect study (Figs. 1 and 2).

The dense reed bed at Lough Avullin. This was an open body of water at the time of the first survey. Water rails *Rallus aquaticus* Linnaeus were present during the 1909-11 survey but have not been recorded since then. Lough Avullin is a habitat that has not been thoroughly surveyed for birds. Photo: T. Kelly

No rails of any kind were detected during the transect survey. However, the moorhen was seen in previous surveys (see Chapter 2 for details) and is believed to have bred on the island in the recent past. Moreover, Lough Avullin, the most likely habitat for both moorhen and water rail was not surveyed in detail in 2013.

Discussion

Transect based surveys that were not specifically focused on waders and wildfowl may not have been optimal for quantifying the numerical abundance of these bird species or for pinpointing their exact geographical distribution. While many wading bird species including common snipe, lapwing, and common sandpiper announce themselves to the observer by both aerial displays and loud calls, these taxa may also remain motionless in the nesting habitat and may not be detectable other than at very close range (see especially Ferguson–Lees *et al.* 2011). Consequently, many nest sites may have remained undetected during the type of transect survey conducted in 2006 and 2013 (see for example Colhoun *et al.* 2015 and references therein). An additional caveat is that in contrast to the 2006 survey, the 2013 transects did not reach areas where, for example, lapwing, and especially ringed plover, had been detected in the past e.g. Beetle Head (Coombes 2003; Winters 2006; Eoin McGreal, pers. comm.). Thus, while a dedicated habitat-specific study would have been preferable, this was not possible during the relatively brief expedition to Clare Island in 2013.

Nevertheless, while Winters (2006) did not record common snipe, these surveys did reveal that the same five species of waders listed by Ussher in 1912 continue to nest on Clare Island and probably have done so over most of the 100-year interval. Thus 50% of the regularly breeding wading bird avifauna of Ireland (including red-necked phalarope *Phalaropus lobatus* (Linnaeus) but excluding woodcock *Scolopax rusticola* Linnaeus) are present on Clare Island. This figure compares with two species (oystercatcher and lapwing) on Great Saltee (89ha), Co. Wexford (Lloyd 1983) and four species (dunlin *Calidris alpina* (Linnaeus), oystercatcher, lapwing and ringed plover) on the Inishkea Islands (360ha), Co. Mayo (Cabot 1963). Common snipe was not recorded on the Inishkea Islands (Cabot 1963) nor on the Aran Islands (4,567ha), Co. Galway, where oystercatcher, lapwing, ringed plover and common sandpiper breed (Lysaght 2002). Interestingly the red-breasted merganser *Mergus serrator* Linnaeus breeds on Great Saltee, the Aran Islands and the Inishkeas (Lloyd 1983; Lysaght 2002; Cabot 1963) but not on Clare Island (Ussher 1912; this study). Additionally, with shelduck *Tadorna tadorna* (Linnaeus) and mallard breeding on Great Saltee, the Inishkeas and the Aran Islands—the latter also with the mute swan *Cygnus olor* (J.F. Gmelin) (Lloyd 1983; Lysaght 2002; Cabot 1963), these archipelagos have a richer anatid avifauna than exists on Clare Island. Therefore, area *per se* does not appear to be a reliable predictor of species richness of breeding water birds on Irish islands.

It is also interesting to note that while Co. Mayo remains a breeding stronghold of lapwing and ringed plover (albeit with a clear decline in numbers over the past 20 years) (Suddaby *et al.* 2010; Balmer *et al.* 2013), other species including dunlin also breed at sites on the coast near Louisburgh and on the Inishkea Islands (Suddaby *et al.* 2010). Therefore, the dunlin breeds at several points along the coast of Mayo but not on Clare Island where there is no machair habitat (Eoin McGreal, pers. comm.). The dunlin breeding population in Ireland, along with that of the curlew *Numenius arquata* (Linnaeus), has declined very steeply in recent years (Olivia Crowe, pers. comm.).

It appears that on Clare Island the numbers of breeding pairs of waders and ducks are low, and these are thinly distributed over the island (Fig. 2). Moreover, this also appears to have been the case during the first Clare Island survey (Ussher 1912) when all the wader species (except common snipe) and mallard were only present in low numbers. This pattern of low numerical abundance of aquatic birds is, on the one hand, somewhat unexpected because the estimated annual precipitation (Vullings *et al.* 2013) is approximately 1,300mm (some 250mm higher than on the island of Ireland as a whole)—a volume of rainfall that would be likely to lead to the formation of open wetland habitats—but, on the other, 'there are surprisingly few waterbodies on Clare Island despite the impervious underlying silicaceous geology' (Ryle 2013). Further, there appear to be even fewer permanent open bodies of water now than there were 100 years ago (Praeger 1911) as shown by Ryle (2013) and Vullings *et al.* (2013) (see also Coxon 2001). And while some areas of open water are small and ephemeral, often associated with peat extraction, only two of the five permanent 'lakes' listed by Praeger (1911) remain today (Ryle 2013). Most notable is the transformation of Lough Avullin from an open body of water of 1.5ha in extent, and up to 2.5m deep in places 100 years ago, to a quaking reed swamp (Coxon 2001; Coxon *et al.* 2013; Ryle 2013; Vullings *et al.* 2013) devoid of lacustrine habitats suitable for breeding waterfowl. There are of course extensive areas of blanket bog on Clare Island (Coxon 2001; Vullings *et al.* 2013; Ryle 2013). This habitat, together with areas of rough grazing, is suitable for breeding common snipe, while areas of closely cropped grassland provide conditions suited for breeding lapwing and ringed plover. Finally, grassy areas above the littoral would provide suitable breeding sites for the oystercatcher.

All of Clare Island's adult breeding wading bird species together with their fully feathered (ptilopaedic) and immediately self-feeding (precocial; nidifugous) chicks are likely to feed on terrestrial invertebrates including earthworms (Lumbricidae), snails (Mollusca), and insects such as beetles (Coleoptera) and fly larvae, e.g. Tipulidae (craneflies), as well as Lepidoptera, and spiders (Arachnida: Araneae) (see Cramp and Simmons 1983; Buchanan *et al.* 2006). The common sandpiper, which favours shingle and gravel areas within rivers and streams, will also take freshwater invertebrates including stoneflies (Plecoptera) and caddis (Trichoptera) as well as other groups (see Cramp and Simmons 1983). The Lumbricidae, Tipulidae, Coleoptera and Araneae as well as snails are all present on Clare Island (the late Dr T.K. McCarthy, pers. comm.). Therefore, it would be difficult to argue that availability of prey is a likely explanation for the low numbers of breeding waders.

Although breeding at low densities, relatively larger wader breeding populations are known on the Mayo and Galway coasts where mammalian predators are more abundant (Suddaby *et al.* 2010), whereas on Clare Island the brown rat *Rattus norvegicus* Berkenhout and domestic and possibly feral cats *Felis catus* Linnaeus are present, but the red fox *Vulpes vulpes* Linnaeus, stoat *Mustela erminea* Linnaeus and American mink *Neovison vison* Shreber are all absent (Colin Lawton in press). Nevertheless, the lapwings on Great Saltee Island, as vividly described by Lloyd (1983), all failed to breed successfully 'mainly because of interference by gulls, crows and rats' but predator exclusion fencing on the Shannon Callows appears to have led to an increase in breeding waders there (Finney and Donaghy 2014). However, gulls, hooded crows *Corvus cornix* Linnaeus, ravens *Corvus corax* Linnaeus (Chapter 2), and rats are all present on Clare Island and together these may account, in addition to possibly poor habitat and resource availability, for the low numbers of breeding waders. In summer, when the eggs hatch, high rainfall accompanied by Atlantic gales, combined with trampling of eggs and chicks by livestock, may be additional hazards.

Excepting ongoing work at the Shannon Callows (Donaghy *et al.* 2014) and previous research in Northern Ireland (e.g. Grant *et al.* 1999; Henderson *et al.* 2002; see also Colhoun *et al.* 2015), the breeding biology of most wading bird species has not been studied in detail in Ireland, and particularly on Clare Island, though basic features such as clutch sizes are well known (Cramp and Simmons 1983). Thus, the common sandpiper, ringed plover, lapwing and common snipe all typically produce clutches of four eggs (with replacement clutches if the first attempt is destroyed), whereas the oystercatcher has a clutch size of three eggs (Cramp and Simmons 1983). Most of the species are single brooded except for the ringed plover, which has two to three broods, and the common snipe, which may produce a second brood. Nevertheless, a relatively high fledging rate is necessary to maintain a stable population, even though adult mortality may be low (e.g. Peach *et al.* 1994).

Stream at Lecarrow where the common sandpiper *Actitis hypoleucos* (Linnaeus) bred in 2013 and where perhaps the dipper *Cinclus cinclus* (Linnaeus) bred in the 1909–1911 interval? Photo: T. Kelly

Ussher (1912) reported that the water rail bred at Lough Avullin during the first survey and that the moorhen nested at some of the other lake sites. However, the water rail has not been recorded, or proven to have bred, on Clare Island since the time of the first survey but the current dense reed beds at Lough Avullin make such a confirmation rather difficult even though the habitat looks very suitable for nesting by this species. In addition to the recording of the water rail as a breeding species, at least five individuals were shot in winter during the first survey (Ussher 1912). This proves that the water rail was relatively common on Clare Island during the 1909–11 period. The moorhen has been seen several times in recent years with evidence that it has bred in the marsh near the beach (see Chapter 2). However, the moorhen was not detected during the recent 2013 transect survey.

The common sandpiper may be the only habitual migrant shorebird on Clare Island as it winters in West Africa south of the Sahara (Wernham *et al.* 2002). All breeding populations in Ireland of the other species may be sedentary (Delany *et al.* 2009), though it appears that very few nestlings have been ringed in this country to prove this point (see Wernham *et al.* 2002). Nevertheless, the breeding waders of Ireland, including those on Clare Island, may be reproductively isolated from other populations in Western Europe, Iceland and Greenland, and as such their genetics would appear to be worth investigating.

While the open water of Lough Avullin on Clare Island has been colonised and replaced by *Phragmites*, the damning of the 'main drain' on Inishmurray (Sligo) created the 'large pool' (Heraughty 1982), which may have enhanced the diversity of aquatic birds including the confirmed breeding of teal *Anas crecca* Linnaeus discovered by Cabot (1962). Rats are absent from Inishmurray (Heraughty 1982). Management of island water bodies may therefore potentially improve the suitability of islands for waders and waterfowl. More generally, offshore islands should be evaluated for use as conservation sites for breeding waders and wildfowl in Ireland; islands generally have relatively few ground predators thus potentially enhancing breeding success among accessible and therefore vulnerable wading bird populations.

Acknowledgements
We are grateful to Ms R. Hutchinson-Kelly for assistance with the fieldwork and preparation of this paper. All the work associated with this research was privately funded.

REFERENCES

Balmer, D.E., Gillings, S., Caffrey, B.J., Swann, R.L., Downie, I.S. and Fuller, R.J. 2013 *Bird Atlas 2007–2011: the breeding and wintering birds of Britain and Ireland*. Thetford. BTO Books.

Brown, D., Wilson, J., Douglas, D., Thompson, P., Foster, S., McCulloch, N., Phillips, J., Stroud, D., Whitehead, S., Crockford, N. and Sheldon, R. 2015 The Eurasian curlew - the most pressing bird conservation priority in the UK? *British Birds* **108**, 660–8.

Buchanan, G.M., Grant, M.C., Sanderson, R.A. and Pearce-Higgins, J.W. 2006 The contribution of invertebrate taxa to moorland bird diets and potential implications for land–use management. *Ibis* **148**, 615–28.

Cabot, D.B. 1962 An ornithological expedition to Inishmurray, Co. Sligo. *Irish Naturalists Journal* **14**, 59–61.

Cabot, D.B. 1963 The breeding birds of the Inishkea Islands, Co. Mayo. *Irish Naturalists' Journal* **14**, 113–5.

Calladine, J., Pakeman, R., Humphreys, E., Huband, S. and Fuller, R.J. 2014 Changes in breeding wader assemblages, vegetation and land use within machair environments over three decades. *Bird Study* **61**, 287–300.

Colhoun, K., Mawhinney, K. and Peach, W.J. 2015 Population estimates and changes in abundance of breeding waders in Northern Ireland up to 2013. *Bird Study* **62**, 394–403.

Coombes, R.H. 2003 Clare Island, Co. Mayo. Birds noted on a 3 day visit – 29/4 to 1/5/2000. Unpublished report.

Coxon, P. 2001 The Quaternary History of Clare Island. In J.R. Graham (ed.) *New Survey of Clare Island*. Volume **2**, 87–112. Dublin. Royal Irish Academy.

Coxon, P., Corcoran., R., Gibson, P. and McCarron, S. 2013 A Holocene pollen diagram from Lough Avullin, Clare Island, Western Ireland. In D. Synnott, (ed.) *New Survey of Clare Island*. Volume **7**, 1–26. Dublin. Royal Irish Academy.

Cramp, S. and Simmons, K.E.L. (eds) 1983 *Handbook of birds of Europe, The Middle East and North Africa*. Vol. 3. Waders to Gulls. Oxford. Oxford University Press.

Delany, S., Scott, D., Dodman, T. and Stroud, D. (eds) 2009 *An Atlas of Wader Populations in Africa and Western Eurasia*. Wageningen. Wetlands International.

Donaghy, A., Prosser, C. and Lynch, A. 2014 Factors affecting breeding success of waders at the Shannon Callows *Irish Birds* **10**, 37–46.

Ferguson-Lees, J., Castell, R. and Leech, D. 2011 *A Field Guide to Monitoring Nests*. Thetford. British Trust For Ornithology.

Finney, K. and Donaghy, A. 2014 The effect of management measures on breeding wader populations in the Shannon Callows *Irish Birds* **10**, 1–10.

Forsyth, I. and Allen, P. 1997 Report to the Royal Irish Academy on the Breeding Birds of Clare Island, Co. Mayo 1996 and 1997. Unpublished report.

Gibbons, D.W., Reid, J.B. and Chapman, R.A. (eds) 1993 *The New Atlas of Breeding Birds in Britain and Ireland: 1988–1991*. London. Poyser.

Grant, M.C., Orsman, C., Easton, J., Lodge, C., Smith, C., Thompson, G., Rodwell, S. and Moore, N. 1999 Breeding success and causes of breeding failure of curlew *Numenius arquata* in Northern Ireland. *Journal of Applied Ecology* **36**, 59–74.

Heath, M.F., Borrgreve, C. and Peet, N. 2000 *European Bird Populations: Estimates and Trends*. Cambridge. Birdlife International.

Henderson, I.G., Wilson, A.M., Steele, D. and Vickery, J.A. 2002 Population estimates, trends and habitat associations of breeding lapwings *Vanellus vanellus*, curlew *Numenius arquata* and snipe *Gallinago gallinago* in Northern Ireland in 1999. *Bird Study* **49**, 17–25.

Heraughty, P. 1982 *Inishmurray ancient monastic island*. Dublin. O'Brien Press.

Jackson, D.B. 2001 Experimental removal of introduced hedgehogs improves wader nest success in the Western isles, Scotland. *Journal of Applied Ecology* **38**, 802–12.

Jackson, D.B., Fuller, R.J. and Campbell, S.T. 2004 Long-term population changes among shorebirds in the Outer Hebrides, Scotland, in relation to introduced hedgehogs (*Erinaceus europeus*). *Biological Conservation* **117**, 151–66.

Jackson, D.B. and Green, R.E. 2000 The importance of the introduced hedgehog (*Erinaceus europeus*) as a predator of the eggs of waders (Charadrii) on Machair, on South Uist, Scotland. *Biological Conservation* **93**, 333–48.

Kelly, T.C. 2013 Unpublished data, Clare Island, Co. Mayo. Unpublished report.

Kirwan, G., Porter, R. and Scott, D. 2015 Chronicle of an extinction? A review of slender-billed curlew records in the Middle East. *British Birds* **108**, 669–82.

Lack, P.C. 1986 *The Atlas of Wintering Birds in Britain and Ireland*. Calton. Poyser.

Lloyd, C.S. 1983 The status of landbirds breeding on Great Saltee, Co. Wexford. *Irish Naturalists' Journal* **21**, 97–103.

Lloyd, C.S. 1984 The birds of Clare Island, Co. Mayo in June 1982. *Irish Naturalists' Journal* **21**, 212–6.

Lysaght, L. 2002 *An Atlas of Breeding Birds of the Burren and the Aran Islands*. Dublin. BirdWatch Ireland.

Macdonald, M.A. and Bolton, M. 2008 Predation of wader nests in Europe. *Ibis* **150** (Suppl. 1), 54–73.

Munro, M. 2017 What's killing the World's shorebirds? *Nature* **541**, 16–20.

Peach, W.J., Thompson, P.S. and Coulson, J.C. 1994 Annual and long term variation in the survival rates of lapwings *Vanellus vanellus Journal of Animal Ecology* **63**, 60–70.

Praeger, R.L. 1911 Clare Island Survey Part 10 Phanerogamia and Pteridophyta. *Proceedings of the Royal Irish Academy* **31**, 1–112.

Ryle, T. 2013 Vegetation-Environment interactions on Clare Island. In D. Synnott (ed.) *New Survey of Clare Island*. Volume **7**, 27–178. Dublin. Royal Irish Academy.

Sandercock, B.K. 2003 Estimation of survival rates for wader populations: a review of mark-recapture methods. *Wader Study Group Bulletin* **100**, 163–74.

Sharrock, J.T.R. 1976 *The Atlas of Breeding Birds in Britain and Ireland*. Tring. British Trust for Ornithology.

Suddaby, D., Nelson, T. and Veldman, J. 2010 Resurvey of breeding wader populations of machair and associated wet grasslands in north-west Ireland. *Irish Wildlife Manuals* No. 44. Dublin. National Parks and Wildlife Service.

Thorup, O. 2006 *Breeding waders in Europe 2000*. International Wader Studies 14. United Kingdom. International Wader Study Group.

Ussher, R.J. 1912 A biological survey of Clare Island in the county of Mayo, Ireland and adjoining district. *Proceedings of the Royal Irish Academy* **31**, Sec. 11, Part 20.

Vullings, W. Collins, J.F. and Smillie, G. 2013 Soils and Soil Associations. *New Survey of Clare Island*. Volume **8**, 1–176. Dublin. Royal Irish Academy.

Wernham, C., Toms, M., Marchant, J., Clark, J., Siriwardina, G. and Baillie, S. (eds) 2002 *The Migration Atlas: Movement of birds of Britain and Ireland*. London. Poyser.

Winters, P. 2006 *A Summer Bird Survey of Clare Island*. Unpublished MSc thesis. Dublin. Centre for the Environment, Trinity College.

CHAPTER 6

BREEDING LAND BIRDS ON THE OFFSHORE ISLANDS OF IRELAND – SOME PATTERNS AND TRENDS

T.C. Kelly, P. Smiddy and U. Keating

ABSTRACT

This study reviews published information on breeding land birds (raptors excluded) of 29 offshore Irish islands. Fifty-seven passerine and near-passerine species have bred on Irish offshore islands. Thus, some 86% of the land bird species breeding on the island of Ireland also breed on offshore islands. A frequency distribution shows that some species occupy all (rock pipit *Anthus petrosus* (Montagu)), or most of the islands (northern wheatear *Oenanthe oenanthe* (Linnaeus)), whereas endemic subspecies (dipper *Cinclus cinclus* (Linnaeus) and coal tit *Periparus ater* (Linnaeus)) have occurred on only one island in the sample. Analysis of the species-area relationship was, unsurprisingly, statistically highly significant. An NMDS ordination analysis also showed area, latitude, the presence of freshwater lentic habitat and human habitations past or present to be significant influences on species richness. An aim of the study was to expand the concept of 'pool size' developed by Russell *et al.* (2006). Thus, the frequency of occurrence of 57 land bird species on the 29 islands (*Y*) was regressed against the overall abundance of their populations on the island of Ireland as a whole (*N*), as well as their total biomass (*M*) ($M = N \times$ bird body weight) and the extent of their geographical distribution (*D*), i.e. the total number of occupied breeding atlas squares. Moreover, based on analysis of the relationship between *N* and *D* on the island of Ireland, and between bird species weight and *D*, it was hypothesised that the number of occupied islands would be correlated with *N* and not with bird weight and therefore, *M*. However, despite the obvious influence of area, the results also show that, against expectation, there is no relationship between *N* and the number of occupied islands (*Y*), whereas there is a relationship between *M* and *Y*. These results suggest that there is a higher proportion of heavier species on offshore islands than on the island of Ireland as a whole, and that those species which are numerically relatively scarce on the 'mainland' occur on a high proportion of offshore islands.

Introduction

Although the Clare Island survey preceded the publication of MacArthur and Wilson's (1967) hallowed monograph on 'The Theory of Island Biogeography' (hereafter IBT), some of the main issues the latter analysed, such as how area as well as isolation from continental land masses affected species richness, had been addressed in the nineteenth century (e.g. Abbott 1980; Connor and McCoy 1979; Rosenzweig and Ziv 1999;

Losos and Ricklefs 2010). However, as explained elsewhere (Chapter 1) these hypotheses were not the principal focus of the original survey, which instead sought evidence in the morphology and biometrics of the birds of Clare Island. In this regard, the original survey focused on endemism, a logical extrapolation from Darwin's (1859) theory of the origin of new species by means of geographical (reproductive) isolation, that is, Allopatric Speciation (e.g. Avise 1987; Avise and Walker 1998; Ridley 2004; Butlin *et al.* 2008; Claramunt *et al.* 2012; see also Collins 1999).

The novelty of MacArthur and Wilson's (1967) monograph related in large part to their formulation of the 'equilibrium theory' (MacArthur and Wilson (1963), which hypothesised that the number of species on an island is a function of its area, and isolation, due to the dynamic processes of immigration and extinction (Losos and Ricklefs 2010). This, in turn, will lead to 'turnover', so while the number of species may remain relatively constant and a function of island area and distance from the continental land mass, the species composition would change over time as new taxa arrive while others become extinct (Whittaker and Fernandez-Palacios 2007; Whittaker *et al.* 2008). The equilibrium number occurs when extinction equals immigration (e.g. MacArthur and Wilson 1963). Further, Russell *et al.* (2006) showed that in the case of a selected list of British and Irish islands, the probability of extinction was lowest on the largest islands and those that are more remote and isolated, i.e. from the nearest 'mainland'. In addition, colonisation was lowest on the more isolated islands (Russell *et al.* 2006).

In the event, Ussher (1912) did not discover any Clare Island endemic forms among the avifauna. However, though the modern technique of mitochondrial DNA (Avise 1987; 1991), which might be able to show those endemic forms, was not, of course, available during Ussher's survey, it could now be applied to the many 'skins' collected then.

The findings of the first and subsequent surveys of the avifauna of Clare Island are however of considerable interest and permit, when combined with additional data on the species composition of other islands in Ireland, an IBT-type analysis. Previously, the species richness of selected islands off the coast of Ireland has been the subject of IBT-type analyses (Reed 1980; 1981; Pimm *et al.* 1993; Russell *et al.* 1995; 2006). The latter study analysed 'pool size' which they defined as 'the number of species with a reasonable chance of reaching the islands' as one of the 'main factors that affect species richness and community dynamics on islands' (Russell *et al.* 2006).

To date, there does not appear to have been a 'global' investigation of the species composition of birds breeding on the islands of Ireland. Therefore, this paper—in addition to quantifying the relationship between bird species richness and island area, distance from the mainland and geographical location—aims to identify the suite of species that most frequently breed on the islands of Ireland and to relate the observed patterns to the species pool. In addition, we amplify the concept of Russell *et al.* (2006); in this context, we take pool size to include not just the species richness but also their numerical abundance on the island of Ireland, their biomasses and the extent of their geographical distribution (number of occupied breeding atlas squares).

Unlike earlier researchers, we are fortunate in that there now exists a detailed data set relating to the above-mentioned variables (e.g. Crowe *et al.* 2010, 2014; Nairn and O'Halloran 2012; Balmer *et al.* 2013; Colhoun and Cummins 2013). The approach here is not to undertake a detailed statistical investigation, but to identify—in a preliminary quantitative survey—the main trends that will be subject to a formal analysis in a subsequent study.

Therefore, this study, in addition to analysing the relationship between land bird species richness and island area, distance from the mainland, geographical position (latitude and longitude), the presence of a lentic habitat above a certain area, whether the island has human habitations and whether a lighthouse is present (Barrington 1900; Stracey and Pimm 2009), also considers the broad relationship between the probability of a species being present on an island and (i) its numerical abundance in Ireland, (ii) its total biomass on the island of Ireland and (iii) the extent of its known geographical distribution (i.e. of atlas squares occupied by each species).

Species pool size: numbers, biomass and the distribution of birds in Ireland

Recent studies by Crowe *et al.* (2010, 2014) are of outstanding interest because they have provided—for the first time—estimates of the numerical abundance of many land bird species in Ireland. These data have been compiled as part of the Countryside Bird Survey (CBS) and the Breeding Bird Survey (BBS) in the Republic of Ireland and Northern Ireland, respectively. Therefore, these databases permit the following questions to be asked. First, are the most abundant bird species in Ireland also those that are most likely to be present on the offshore islands? Second, are those species with the greatest biomass in Ireland also the most likely to be breeding on the offshore islands? Third, are those species with the widest geographical distribution in Ireland also the most likely to be present on the offshore islands?

Of course, other factors also influence the likelihood of a particular species being present on an island, including its area, geographical position (Reed 1980, 1981; Whittaker and Fernandez-Palacios 2007; Losos and Ricklefs 2010) as well as its history of human occupancy, including perhaps the existence of a lighthouse (Barrington 1900; Alexander 1954; Munns 1956; Hounsome and Rear 1966). The area of an island (and the presence or absence of human settlements) is also related to the variety of habitats it holds (and, by inference, its bird species richness); this concept has always been to the forefront in the hypotheses of biogeographers (e.g. Reed 1981; Lloyd 1984) but, following the work of Simberloff (1976), whose experimental results did not support the expected relationship, the idea has proven difficult to sustain, despite its strong intuitive appeal (e.g. McGuinness 1984). The most likely explanation for this is that there are special difficulties associated with defining and quantifying the extent of different habitats for breeding birds—other than on the broadest scale—including those on islands (e.g. see Lysaght 2002; Nairn and O'Halloran 2012; Fuller 2012).

Species pool size: numbers of land bird species breeding in Ireland

It is reasonable to hypothesise that the numerical abundance of different bird species on an adjacent landmass enhances the probability that the commoner taxa are more likely to be found on the offshore islands rather than those that are rare. Fig. 1 shows the estimated mean numerical abundance of various land bird species in Ireland based on the data presented in Crowe *et al.* (2014)—and for those taxa not included in the CBS-BBS surveys—derived from the estimates in Gibbons *et al.* (1993).

It can be seen in Fig. 1 that among the top ten most abundant species are the wren *Troglodytes troglodytes* (Linnaeus), barn swallow *Hirundo rustica* Linnaeus, robin *Erithacus rubecula* (Linnaeus) and blackbird *Turdus merula* Linnaeus, but also the rook *Corvus frugilegus* Linnaeus, jackdaw *Corvus monedula* Linnaeus and woodpigeon *Columba palumbus* Linnaeus. While it is acknowledged that these data have been compiled relatively recently (since 1998 in the case of CBS and 1994 in the case of BBS) and that much of the information on the island avifaunas is of a different vintage (mainly 1950 to 2000), there is no objective reason to believe that, for example, a CBS in Ireland in the 1950s would have yielded a very different set of results, though the woodpigeon is much more abundant now than it was in the past (e.g. O'Regan *et al.* 2012), possibly due to the effects of climate change (e.g. Balmer *et al.* 2013). Nevertheless, in terms of the order of relative abundance there is reasonably close agreement between the results of Crowe *et al.* (2014)—which are based on detailed statistics derived from a carefully planned survey methodology—and those of Gibbons *et al.* (1993), which are derived from a much broader and less quantitative approach.

Species pool size: biomass of land bird species breeding in Ireland

These observations are relevant when considering Fig. 2, which is a compilation of the estimated biomass of land bird species breeding in Ireland. The data in Fig. 2 were obtained by multiplying the estimated mean number of individuals of each species breeding in Ireland by their mean weight in grams. The latter was obtained by summing the mean weights of the male and female (Dunning 2008; BWP*i* V2) for each species and dividing by two. Therefore, while the robin, swallow and blackbird again appear in the top ten in terms of their biomasses, that list also includes the rook, hooded crow *Corvus cornix* Linnaeus and magpie *Pica pica* (Linnaeus).

The relationship between numerical abundance and size is elegantly summarised by Lawton (1989) as 'per capita use of resources is less in small-bodied individuals than in large ones; hence finite food resources will support more small individuals than large ones'. In general, larger birds

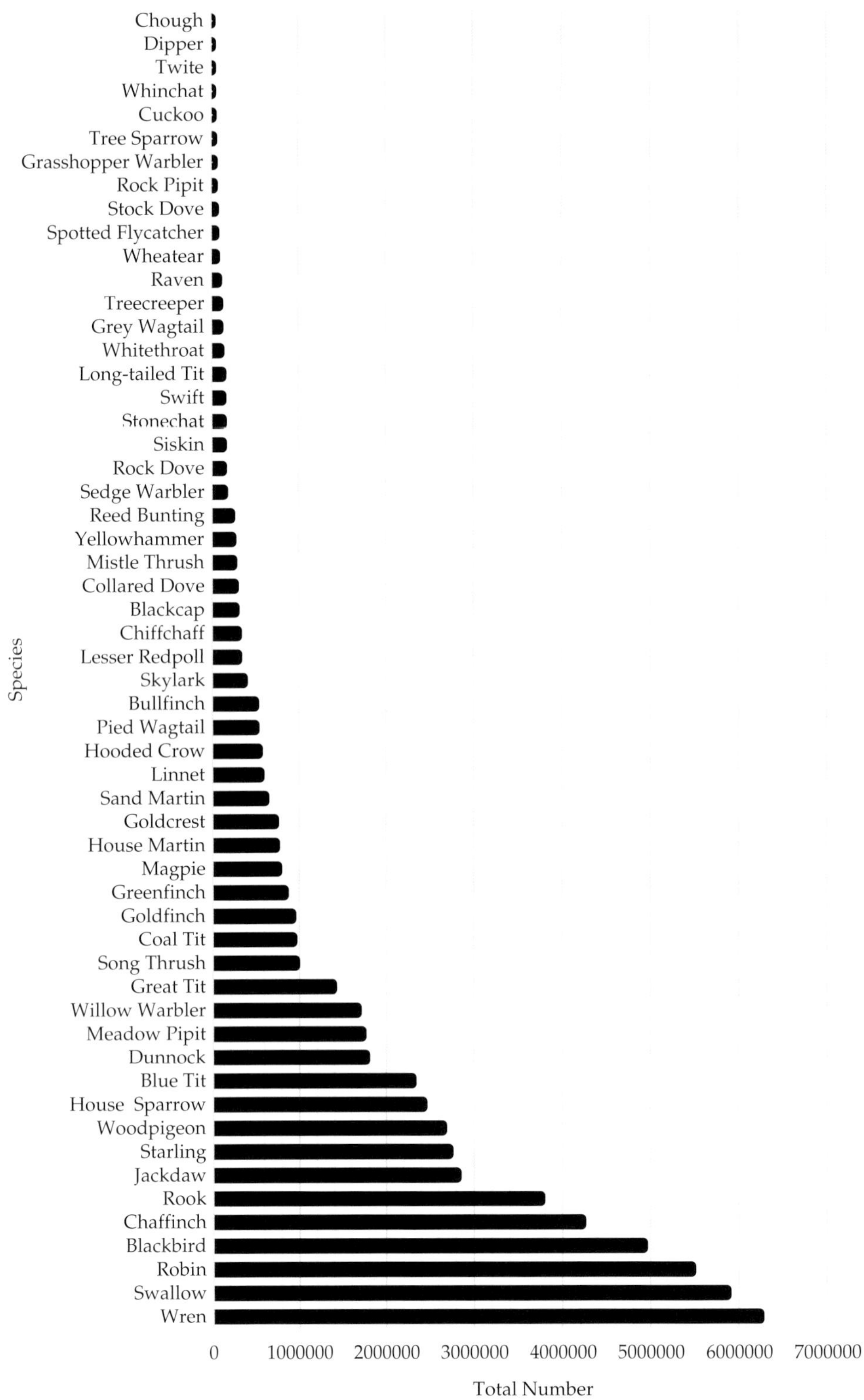

Fig. 1 Number of individuals of each species in Ireland; several species omitted including pheasant, all waders, raptors, and seabirds (modified from Crowe *et al.* 2014).

consume larger amounts of food be it invertebrate prey or vegetation. Larger birds, especially in the temperate region, have higher basal metabolic rates (BMR) (e.g. Williams *et al.* 2010; McKechnie 2015) and so the number of species with a relatively high biomass on islands would appear to be worth investigating. However, the apparently simple relationship depicted by Lawton (1989) is, as he went on to show, somewhat deceptive and it appears that widely accepted explanations of the interactions between biomass, abundance and geographical range size have been

Fig. 2 Biomass (in kg) of land birds in Ireland (see Introduction and Methods).

difficult to prove (e.g. Lawton 1989; Blackburn *et al.* 1993, 1997; 2006; Gaston 1994; Marquet *et al.* 1995; Gaston *et al.* 2000; Storch and Gaston 2004; Reuman *et al.* 2009; Sutherland *et al.* 2013). Despite this difficulty it seems reasonable to hypothesise that land birds with a large mass should have difficulty in colonising islands, whereas smaller species, with a low body mass, might display the opposite trend. Consequently, an inverse relationship between bird weights (and their biomass, i.e. *N**weight) and the number of islands they occupy would be expected; by this reckoning large birds should either be absent or at least very rare on the offshore islands.

Therefore, in addition to the issues of abundance and biomass, it is worthwhile to explore the influence of the width of the geographical distribution of bird species in Ireland, i.e. its 'area of occupancy' (Gaston 1994) on the likelihood that it will colonise an island (see Appendix 1*)—though this will also depend on the size of the island and the variety of habitats it holds.

Aims

The aims of this study are (i) to describe the community of passerine and near-passerine species richness on 29 islands off the coast of Ireland and (ii) to relate the species richness of these land birds to the area of the island. In addition, we undertake a preliminary exploration of the relationships between the patterns of occupancy by birds (number of islands inhabited by the various species) of the sample of Irish islands and their total numerical abundance *N*, biomass *M* and geographical distribution *D*. Finally, we also investigate other variables including latitude and longitude, the distance from the nearest mainland, presence or absence of lentic habitats, current and or past human habitations including the presence of lighthouses on the islands. The latter variable was considered because island lighthouses attract migrating birds (Barrington 1900) and this may, in theory, enhance species richness (Stracey and Pimm 2009).

Methods

The sample of islands

Historically, breeding bird surveys have been conducted on several islands off the coast of Ireland. However, in this paper we have excluded those that are very close to the mainland (≤1km) such as Sherkin, Valentia and Achill islands. It is likely that the nearby mainland, for at least some land bird species on these islands, is part of their territory and within which they routinely forage. Consequently, this gives a sample of 29 islands (including Clare Island) located at different points around the coast of Ireland, but mostly in the west where there are more islands (Fig. 3). The area of each island in the sample, the distance from the mainland, the latitude and longitude, the presence or absence of freshwater lentic habitats and the presence or absence of human occupancy current and historical, is given in Appendix 2*.

Species groups: land birds

As mentioned, in addition to our aim of undertaking a preliminary IBT analysis, we have also attempted to compile a complete list of the bird species breeding on the islands of Ireland. However, in line with the bulk of IBT literature, most of our analyses focus on the land bird fauna of the islands of Ireland. Thus, all the analyses are confined to the passerine and near-passerine species. It should be noted however, that because of issues with the reliability of breeding records of some migrant species in the offshore island bird assemblage—notably, garden warbler *Sylvia borin* (Boddaert) and ring ouzel *Turdus torquatus* Linnaeus—these were excluded from the Non-metric multidimensional scaling (NMDS) statistical analysis. In addition, the great spotted woodpecker *Dendrocopos major* (Linnaeus) was also omitted because of the recency of its breeding in Northern Ireland in 2006 and subsequently in the Republic of Ireland (Hillis 2010; see also Kelly 2008). Other groups such as raptors and occasionally shorebirds will be mentioned but these will not be analysed in detail.

Species pool size

In addition to the number of land bird species breeding on each of the islands in our sample, we have also attempted an analysis based on a broader definition of the 'pool size' concept. Thus, we depict the pool as consisting of *N*, the numerical abundance of the taxa; *M*, the biomass of the land bird species; and *D*, their distribution on the 'mainland' of Ireland (e.g. Passy 2012). The hypothesis is that individually (or in combination) *N*, *M* or *D* might explain the pattern of distribution of the land bird species on the islands of Ireland. We have chosen the whole island of Ireland as the 'pool' of species because unlike Great Britain there is no major north/south variation in species type or richness; in terms of these two categories the island of Ireland is much more homogenous than is Britain (e.g. Balmer *et al.* 2013).

Species pool size: numerical abundance

The data are derived almost exclusively from Crowe *et al.* (2014), but four species, namely

* Appendix 1 can be found on pages 132–35. Appendix 2 can be found on page 137.

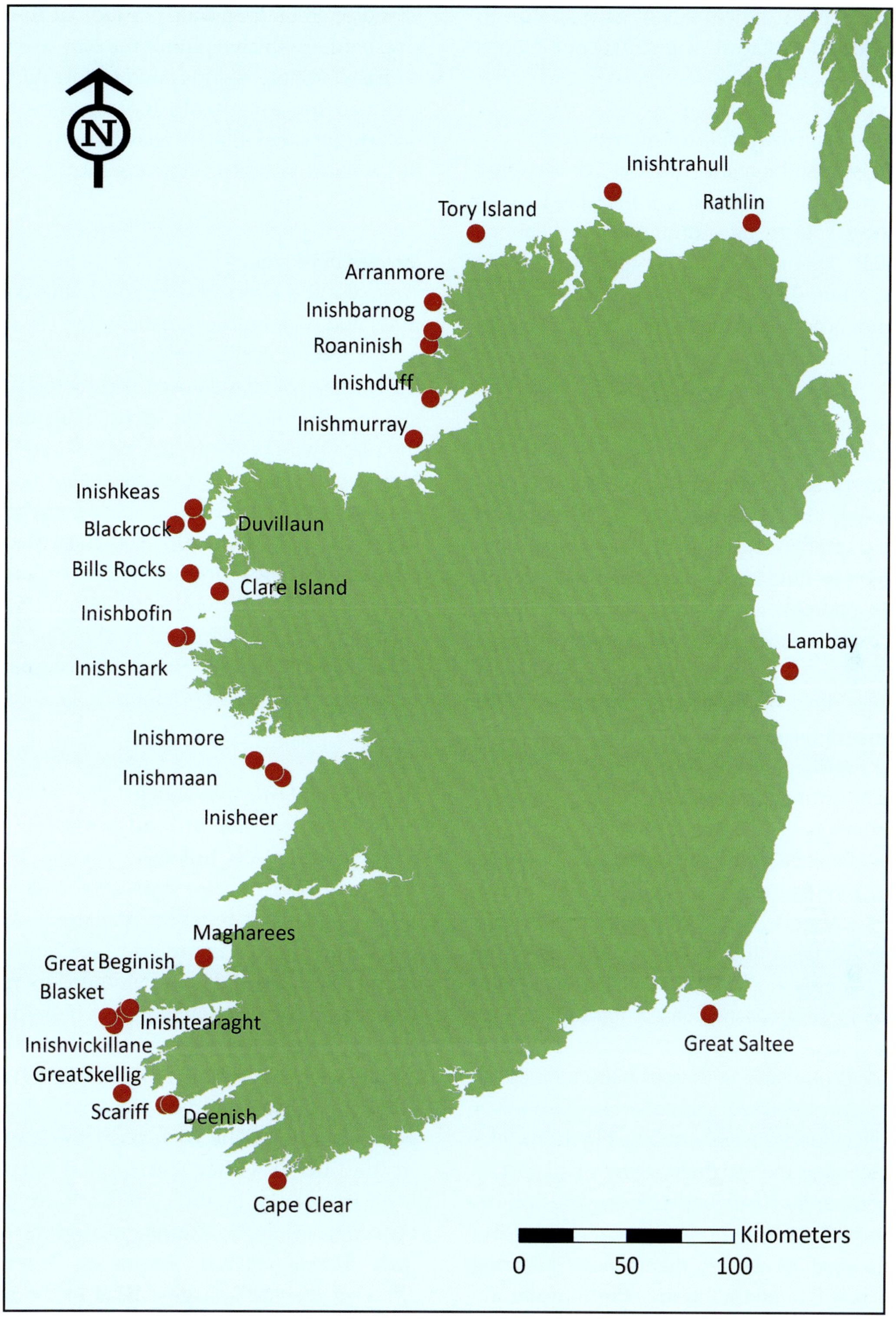

Fig. 3 The locations of the sample of 29 offshore islands investigated in this study.

chough *Pyrrhocorax pyrrhocorax* (Linnaeus), rock dove *Columba livia* Gmelin, rock pipit *Anthus petrosus* (Montagu) and tree sparrow *Passer montanus* (Linnaeus) were not included in that study, so the data presented in Gibbons *et al*. (1993) was used instead.

Species pool size: biomass

The biomass of individual species follows methods outlined in Chapter 5, i.e. by averaging the weights of the male and female of each species (where available) based on data in Dunning (2008) and BWP*i* v2. This figure is then multiplied by the average number of individuals of each species in Ireland reported by Crowe *et al.* (2014) or Gibbons *et al.* (1993), as appropriate.

Species pool size: geographical distribution

The data is based on the number of occupied breeding atlas squares for each of the island group of land bird species as summarised in Gibbons *et al.* (1993). This database was compiled before the recent interval of climate change is believed to have commenced in Ireland in the early 1990s (e.g. Woodward *et al.* 2010).

Statistical analysis

IBT analysis

A preliminary conventional IBT analysis was performed using the familiar Log_{10} Area regressed against Log_{10} Species Richness to assess the similarity between breeding bird communities in the study islands. In addition, non-metric multidimensional scaling (NMDS) ordination analysis with Sørensen distance measures was undertaken to produce ordination plots with an associated dendrogram. NMDS ordination and the dendrogram were created using the Vegan package in R (Oksanen *et al.* 2016). The NMDS ordination was used to investigate if the combination of selected explanatory variables (i.e. area, distance from the mainland, latitude, longitude, past human habitation, present human habitation, presence of a lighthouse, presence of freshwater bodies >100 × 100m) influenced the observed patterns in island bird community assemblages and to group islands based on the similarity of their avifauna. Sørensen distance measurements were used to plot sites in the ordination, based on weighted scores derived from species presence or absence data. Species were positioned in the ordination by obtaining the weighted scores of all species over all sites, and visually representing the optimal conditions for species in NMDS space. The dendrogram was used to observe hierarchical clustering of islands based on the similarity of bird communities according to Sørensen distances. Only species that occurred on 5% or more of the islands were included in the analysis and for the creation of the dendrogram as it is recommended that rarer species are removed when calculating Sørensen distances.

Refining the hypotheses

To refine our hypotheses more clearly we decided to look at the relationship between *N* and *D* on the island of Ireland and also to look at the relationship between the weights of the bird species found on Irish offshore islands and *D*. Note that we have confined this analysis to those species that have occurred on the offshore islands of Ireland (Fig. 4) but excluding those rare migrants as outlined in the text.

Species pool size

To assess the effect of numerical abundance, biomass and geographical distribution (*N*, *M* and *D*) each variable was individually regressed against the number of islands occupied by the individual species. To linearise the trend both axes on all the plots were of the form Log_{10} (*X*) by Log_{10} (*Y*). However, because area exerts such a major influence on species richness, a second analysis of the influence of *N*, *M* and *D* was undertaken that was confined to the six largest islands, i.e. those having an area >900ha: Rathlin, Clare Island, Arranmore, Inishmore, Inishmaan and Inishbofin. Again, the *Y* axis was Log_{10} the number of occupied islands versus the Log_{10} of *N*, *M* and *D*.

Results

Island land bird community

Fifty-seven species of land birds (Fig. 4) have been recorded as breeding on the 29 islands analysed here; this represents 86% of the total number of breeding passerine and near-passerine species in Ireland (see Smiddy 2010; Balmer *et al.* 2013). However, there are major differences in the frequency with which the various species have been recorded on the 29 occupied islands (Fig. 4). Notably, the rock pipit occupies 100% of those islands where land birds have been recorded. The rock pipit feeds efficiently in the littoral. Other species that occur with a high frequency include the northern wheatear *Oenanthe oenanthe* (Linnaeus), wren, and skylark *Alauda arvensis* Linnaeus. Interestingly, some of Ireland's largest land birds, including the raven *Corvus corax* Linnaeus, chough and hooded crow have been found on 20 (69%) of the 29 occupied islands (Fig. 4). These species are well known to occupy the coastal zone with the chough being a semi-colonial inhabitant of the Irish coastline from Waterford to Donegal (Gibbons *et al.* 1993; Balmer *et al.* 2013), though it is also found further east on Rathlin Island.

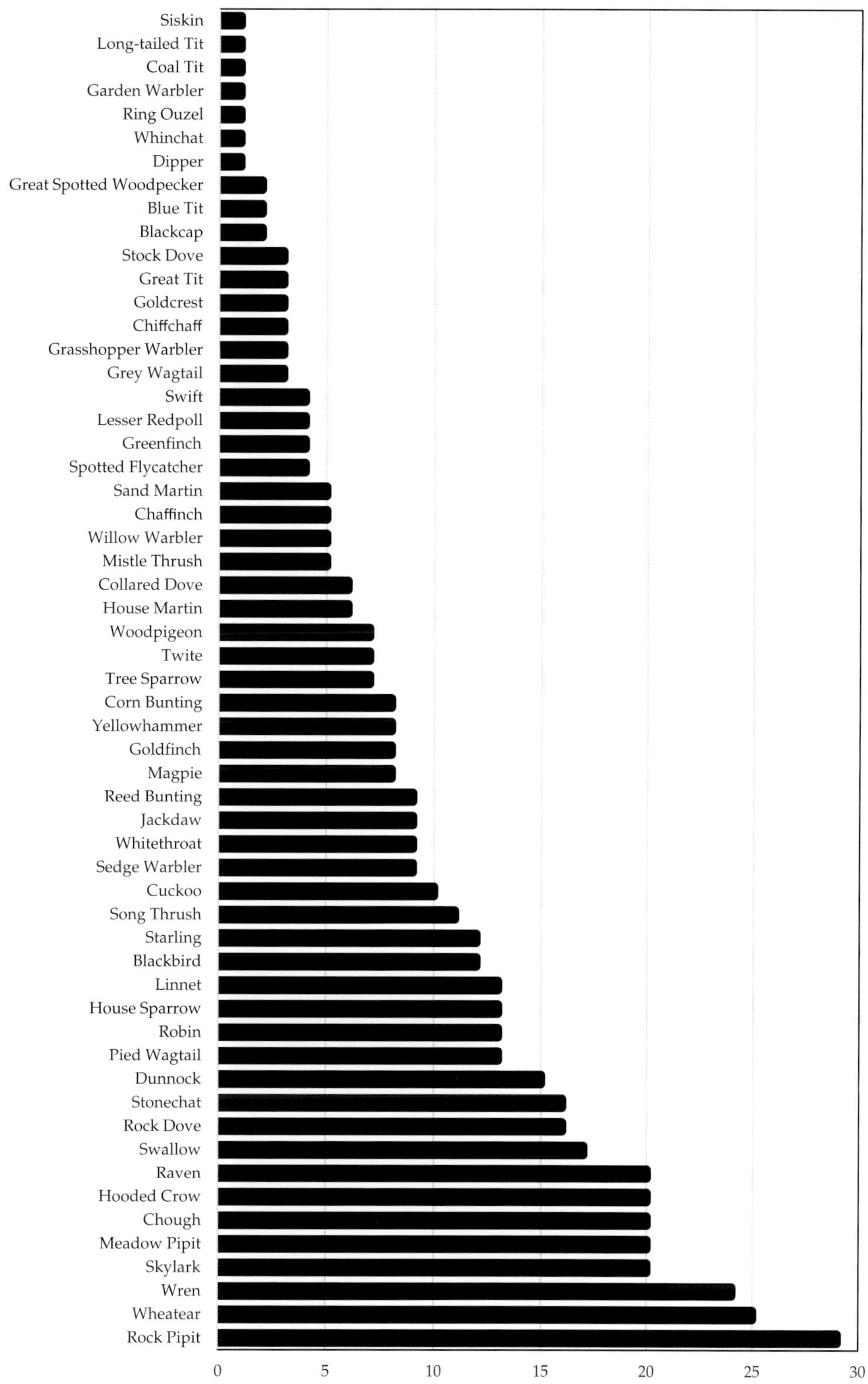

Fig. 4 Frequency distribution of the number of offshore islands (n = 29) occupied by the different species recorded as breeding.

Of the 57 land bird species on the offshore islands, 41 are believed to be sedentary and 15 are migrants (26.8%). Among the latter, the wheatear has been found on 25 Irish islands (86%) and the swallow on 17 (58.6%), though other familiar migrants such as the willow warbler *Phylloscopus trochilus* (Linnaeus) only occur on five (17%) and the common chiffchaff *Phylloscopus collybita* (Vieillot) on three (10%).

Seven raptor species including the white-tailed *Haliaeetus albicilla* (Linnaeus) and golden eagles *Aquila chrysoetos* (Linnaeus) have been recorded as breeding on the offshore islands. The peregrine falcon *Falco peregrinus* Tunstall (n = 13 islands; 44.8%) and common kestrel *Falco tinnunculus* Linnaeus (n = 11; 38%) occur on quite a high proportion of islands and before it became extinct the white-tailed eagle had eyries on seven islands (24%).

The species area relationship

There is, as would be expected, a highly significant correlation between land bird species richness and the area of the islands in the sample. The slope of the line at 0.547 is high but not exceptionally so for island bird assemblages (Fig. 5).

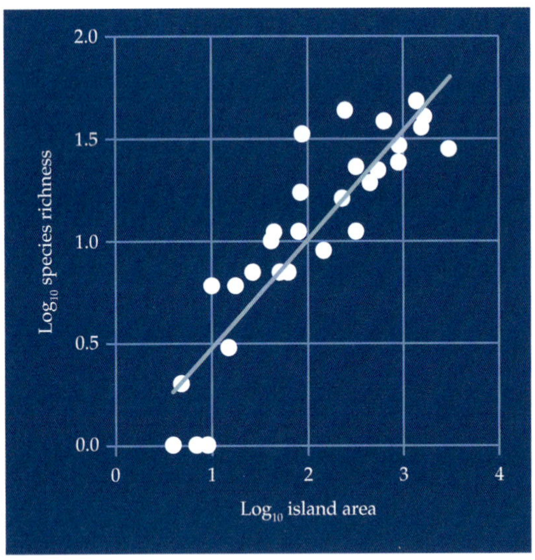

Fig. 5 Log_{10} land bird species richness versus Log_{10} island area relationship ($P < 0.0001$).

Non-metric multidimensional scaling (NMDS) ordination and hierarchical clustering study

The NMDS analysis showed again that area is a major influence on offshore island land bird species richness along with a set of other auto-correlated variables. Thus, large islands will probably have sizeable lentic waterbodies and current or past human habitations. Latitude is also important but not longitude or distance from the nearest 'mainland'. Finally, the presence of lighthouses (Barrington 1900; Stracey and Pimm 2009) does not appear to enhance species richness (Table 1; Fig. 6).

The preliminary Sørensen dendrogram (see Methods) of the similarities of land bird assemblages in our sample of 29 offshore islands shows two main clusters. The first places Tory Island, Inishmurray and Great Blasket Island in one subgroup and all the Aran Islands in another. Arranmore and Rathlin Island are aligned together as are Cape Clear (County Cork) and Lambay Island (County Dublin). Clare Island and Inishbofin form another subgroup (Fig. 7).

The second cluster also includes a diverse group of islands including the geographically widely separated Inishkea and Magharee archipelagos. Inishtrahull (Donegal) is linked with Inishvickillaun (Kerry) and Inishtearaght is clustered with the nearby Great Skellig as well as Scariff and Deenish (Fig. 7).

Table 1 shows the output of the NMDS model. It can be seen that area, latitude, presence of a lentic water body ≥100m^2 and human habitations past or present are strongly associated with

Table 1

Output from the NMDS model showing island area, freshwater lentic habitats, human habitations in the past or present and latitude to be factors significantly (P < 0.05) associated with explaining offshore island land bird species richness (* P = <0.05; ** P = <0.01; *** P = <0.001).

Explanatory variable	NMDS1 | NMDS2	r2	P
Island area	–0.89359 | 0.44889	0.2986	0.010 *
Distance from the mainland	–0.00121 | –1.00000	0.0257	0.701
Presence of freshwater lentic habitat (>100m × 100m)	–0.86531 | 0.50123	0.3819	0.005 **
Past human occupation	–0.89790 | 0.44019	0.6312	0.001 ***
Present human occupation	–0.91996 | 0.39202	0.4621	0.002 **
Lighthouse presence	0.99955 | 0.02989	0.0258	0.697
Latitude	0.05118 | 0.99869	0.2473	0.039 *
Longitude	–0.47705 | 0.87888	0.1137	0.212

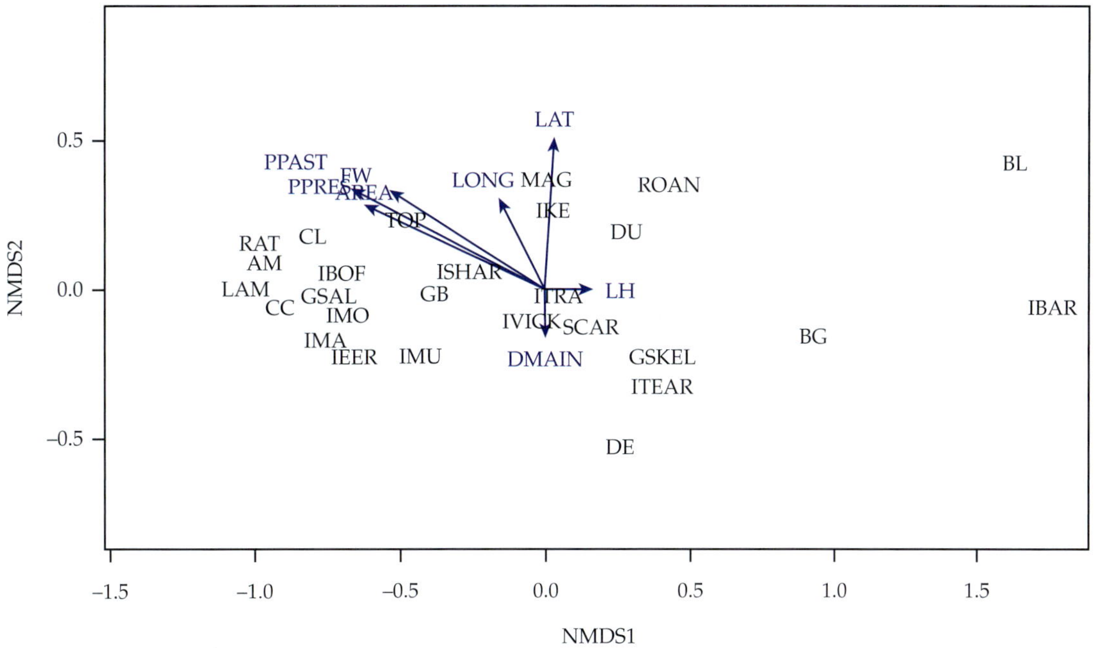

Fig. 6 NMDS ordination of bird community data of species present at >5% of Irish offshore islands investigated in this study. Arranmore (AM); Beginish (BG); Bills Rock (BI); Black Rock (BL); Cape Clear (CC); Clare (CL); Deenish (DE); Duvillaun More (DU); Great Blasket (GB); Great Saltee (GSAL); Great Skellig (GSKEL); Inishbarnog (IBAR); Inishbofin (IBOF); Inishduff (ID); Inisheer (IEER); Inishkeas (IKE); Inishmaan (IMA); Inishmore (IMO); Inishmurray (IMU); Inishshark (ISHAR); Inishtearaght (ITEAR); Inishtrahull (ITRA); Inishvickillaun (IVICK); Lambay (LAM); Magharees (MAG); Rathlin (RAT); Roaninish (ROAN); Scariff (SCAR); Tory (TOR). Vectors of explanatory variables are also shown (the length of the arrow is proportional to the correlation between the variable and the ordination); AREA (island area); DMAIN (distance from mainland); FW (presence of freshwater lentic habitat >100m × 100m); PPAST (past human occupation); PPRES (present human occupation); LH (lighthouse presence); LAT (latitude); LONG (longitude). See Table 1.

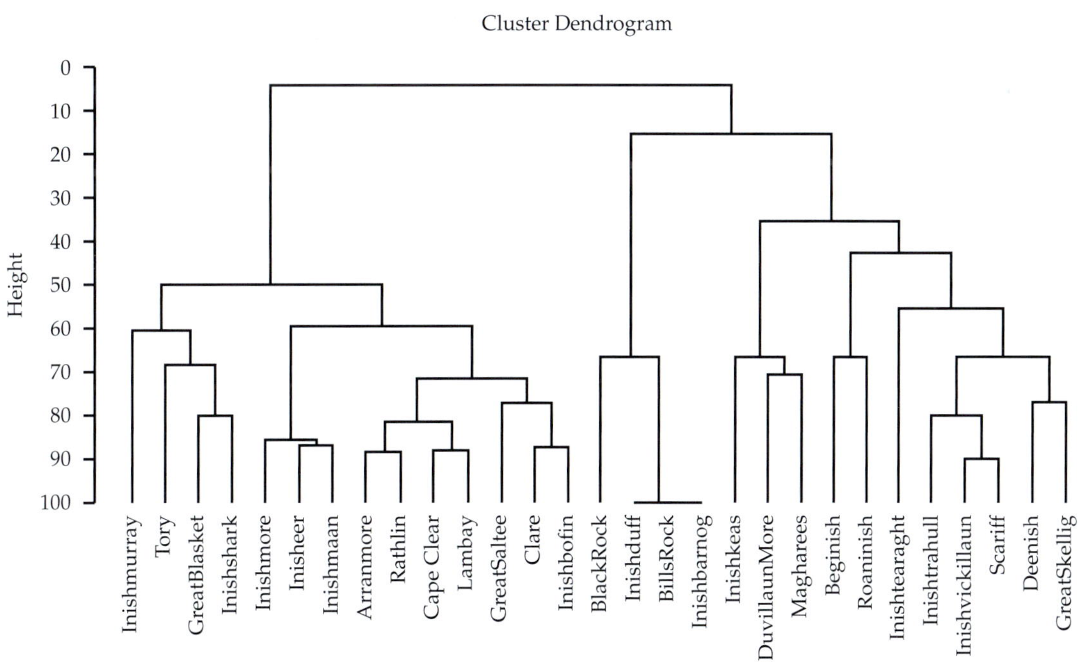

Fig. 7 Hierarchical clustering dendrogram of island community similarities based on Sørensen distances.

the species richness of land birds on the 29 Irish offshore islands. Note that those species with a frequency of 5% or less on the island sample were omitted from the analysis.

Species pool size hypotheses

The relationship between N and D for the island of Ireland is a highly significant positive trend, therefore, we expected a broadly similar relationship between N and Y = the number of occupied islands. In contrast there was no correlation between the weights of land bird species in Ireland and D. Thus, we expected that bird weight and biomass (M) would not exert an influence one way or the other on Y; or should there be a relationship it would be an inverse one.

Species pool size N: all islands

In this study we explored an expanded concept of the 'pool size' developed by Russell *et al.* (2006). Thus, we set out to establish the possible relationship between the numerical abundance of bird species on the island of Ireland (Crowe *et al.* 2014) and the number of offshore islands occupied by those species. As stated above, our hypothesis is that there would be a strong positive correlation between N and Y.

The relationship was tested using $\text{Log}_{10}\ N$ versus Log_{10} Number of occupied islands i.e. Y (n = 29). Although the trend was positive the correlation was not found to be statistically significant ($P = 0.2611$), therefore, our hypothesis regarding the relationship between N and Y was rejected (Fig. 8a).

Species pool size M: all islands

We also explored a possible interaction between the numbers of islands occupied by a species and its body weight, as well as its total biomass, on the island of Ireland. Thus Log_{10} body weight and Log_{10} Biomass, which in turn = $N*M$ were regressed separately against Y (i.e. Log_{10} of number of occupied islands). In both cases (Log_{10} Biomass $P = 0.0205$; body weight $P = 0.019$) the correlations were positive and statistically significant. Consequently, our second hypothesis that there is an inverse correlation with biomass, or body weight (or no influence as is the case with the island of Ireland as a whole) on the frequency of occupation by land bird species, was rejected (Fig. 8b, c). Therefore, M and species weight both

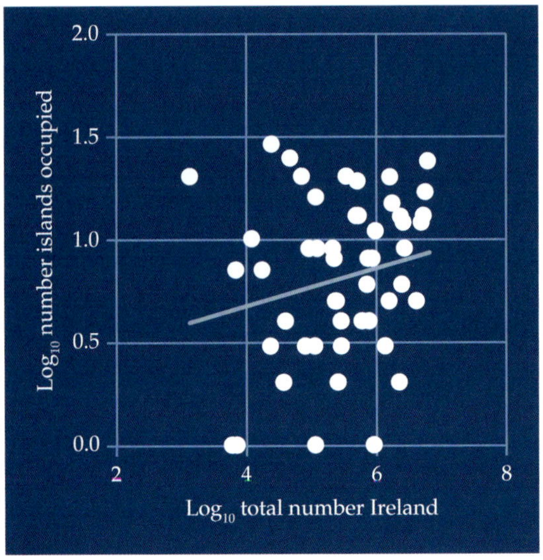

Fig. 8a Relationship between N = Log_{10} total number of land birds of each species in Ireland and Y = Log_{10} the number of islands on which they have been recorded to be breeding ($P = 0.261$).

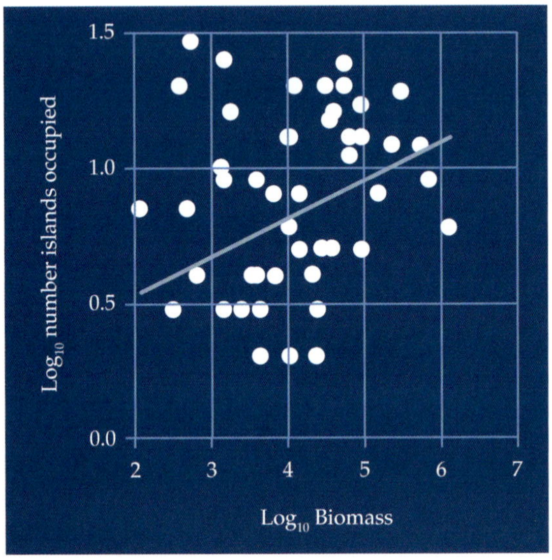

Fig. 8b Relationship between M = Log_{10} biomass and the number of occupied islands ($P < 0.02$).

contribute positively to the likelihood of a bird species being present on an offshore island.

Species pool size D: all islands

The extent of the geographical spread of land bird species breeding on the island of Ireland may influence the likelihood of some or all successfully establishing themselves on an island. The geographical distribution was estimated by including the number of occupied 10km squares in the *New atlas of breeding birds* (Gibbons *et al.* 1993). Thus, Log_{10} Number of occupied breeding atlas squares, i.e. D was regressed against the Log_{10} of the number of islands on which the species was found to be breeding. In this case the relationship, though positive, fell well short of statistical significance (P = <0.107) (Fig. 8d).

Species pool size: larger islands only

As there is a strong correlation between species richness and island area, it is possible the first set of 'pool of species' regressions was ill-conditioned because the number of islands being colonised would be unequally dependent on their areas and therefore obscuring the influence of *N*, *M* and *D*. Therefore, to 'control' for the influence of area we undertook a separate analysis that was confined to the six largest islands in the sample of 29 (see Methods). The first regression of Log_{10} *N* versus Log_{10} Number of occupied islands was not statistically significant (P = 0.266). However, the correlation between *M* (i.e. *N**Biomass) and the number of occupied islands was significant (P <0.029) as was that with body weight (P <0.0345). Finally, while the correlation with *D* (number of breeding atlas squares) was borderline significant (P = 0.0514) the trend, as with all these regressions, was positive (Fig. 9a–d).

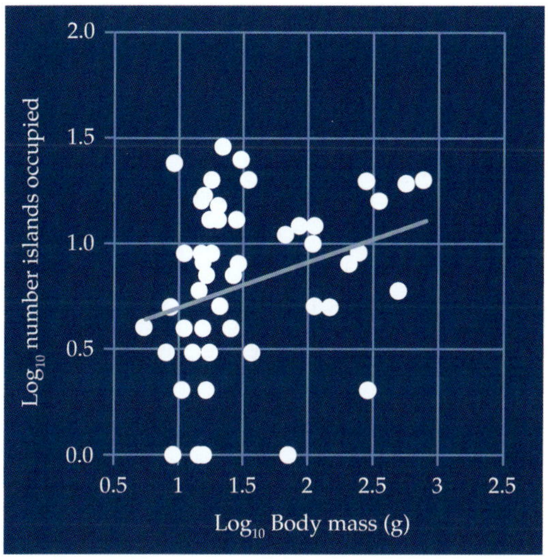

Fig. 8c Relationship between Log_{10} land bird species mass (gm) and Log_{10} number of occupied islands (P <0.019).

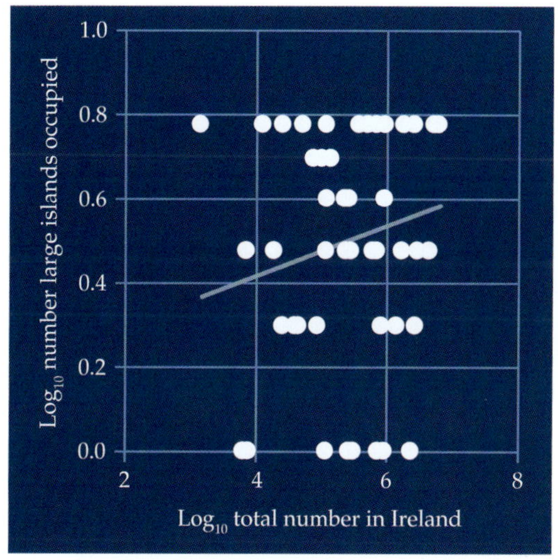

Fig. 9a Relationship between $N = \text{Log}_{10}$ total number of individuals of land bird species in Ireland and Log_{10} number of occupied large islands (P = 0.1927).

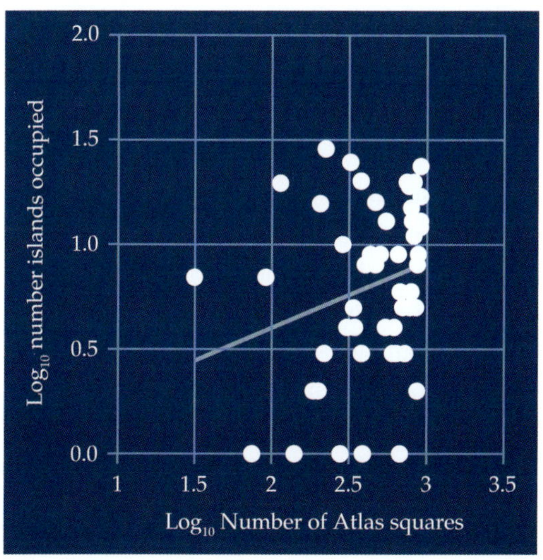

Fig. 8d Relationship between $D = \text{Log}_{10}$ number of occupied breeding atlas squares in Ireland and $Y = \text{Log}_{10}$ number of occupied islands (P <0.107).

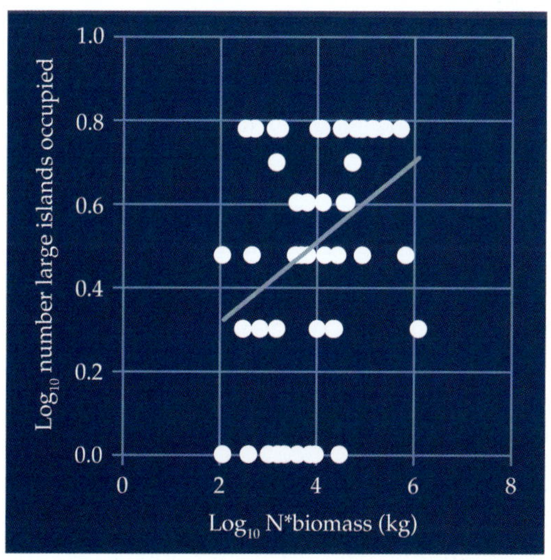

Fig. 9b Relationship between $M = \text{Log}_{10}$ *N** biomass and the number of occupied large islands (P <0.018).

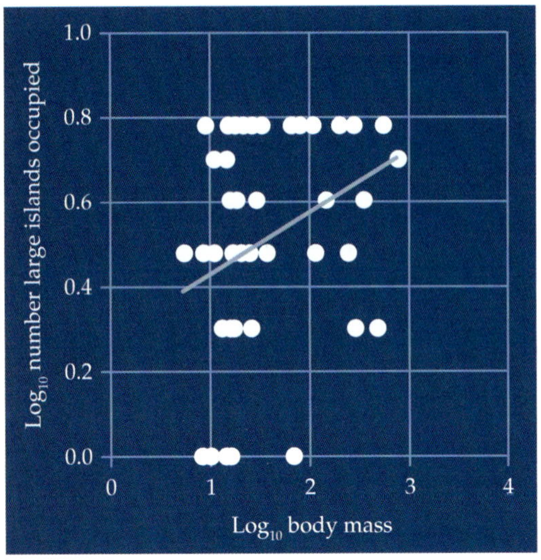

Fig. 9c Relationship between Log_{10} bird species weights (gm) and Log_{10} number of occupied large islands ($P < 0.0395$).

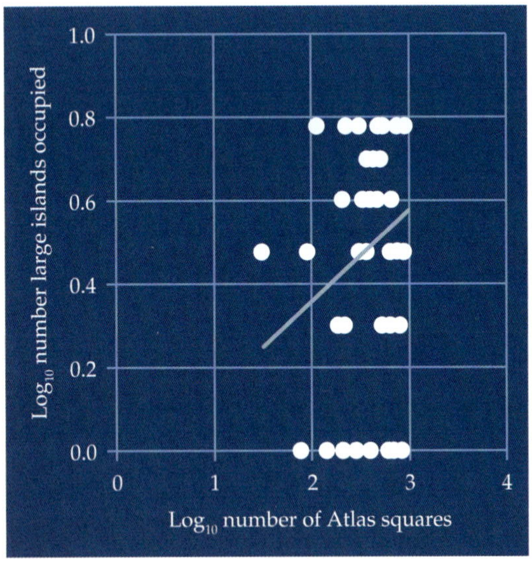

Fig. 9d Relationship between $D = Log_{10}$ number of occupied breeding atlas squares in Ireland and Log_{10} number of occupied large islands $P < 0.0712$.

Discussion

Although the presence of land birds on Irish islands has been touched upon by researchers in the past (see especially Russell *et al.* 2006 and references therein) this is, to our knowledge, the first attempt at delineating the 'assemblage' of species that frequent these marine offshore isles. It is clear however, that our data base is 'ill-conditioned' in the sense that it spans a very long interval of time, stretching back to the period when the white-tailed eagle occupied eyries on seven of our offshore islands. Over the interval of the first Clare Island survey (Ussher 1912) from 1909 to the present, the corn bunting *Emberiza calandra* (Linnaeus) has become extinct and the status and distribution of many other species in Ireland, including the corncrake *Crex crex* (Linnaeus) and twite *Carduelis flavirostris* (Linnaeus), has changed dramatically; in the case of the latter two species involving very marked population declines and range contractions (e.g. Hutchinson 1989; Gibbons *et al.* 1993; Balmer *et al.* 2013). It is also possible that some of the breeding records in the sample of 29 islands fell short of confirmed 'proof'.

In addition, we used the total number of breeding bird species on each island, but this can be criticised on the basis that some islands, such as Great Saltee and Cape Clear, have been surveyed many times whereas others have only been visited once. But while the mean number of species may not vary greatly over succeeding visits the total accumulated number may greatly exceed the average (see McCollin (2014) and the associated Pangaea data set). However, while this procedure of pooling data from infrequently surveyed islands with those subject to near annual investigation appears to be widely practiced, the 'ideal' would be to analyse the equilibrium number of species, but this is often difficult to estimate (e.g. Diamond and May 1977).

Moreover, proving that a bird species is breeding is often difficult and while its presence has been accurately detected, time may not be available, particularly on short visits typically associated with fieldwork on islands, to obtain the confirmatory evidence (see Balmer *et al.* 2013, Table 2.2, page 42). For example, the chiffchaff has been detected during the breeding season on several occasions on Cape Clear Island; singing birds have been present but proof of breeding could not be obtained despite the daily presence of experienced fieldworkers (e.g. Boyle 1997). The difficulty is particularly acute in relation to migrant bird species which, as many ornithologists will have experienced, sing soon after their arrival (e.g. grasshopper *Locustella naevia* (Boddaert), willow and sedge warbler *Acrocephalus schoenobaenus* (Linnaeus)), but depart from these locations (e.g. Great Saltee Island (Wexford); Old Head of Kinsale (Cork); T.C. Kelly unpublished data) within a few days or even hours.

In addition, the spring migration may often extend into the middle of June or later, and thus while an accurate identification may have been made the record may be a 'false positive' as far as breeding confirmation is concerned. Therefore, the single islands on which the whinchat *Saxicola rubetra* (Linnaeus), garden warbler and ring ouzel have occurred should be treated with some caution. In fact, the ring ouzel appears to be an error in the reading of Ussher's paper; this species never bred on Clare Island and does not appear to have ever been reliably recorded there! It should be noted again that all these species were omitted from the detailed analysis.

On the other hand, if a species has been recorded as breeding on an island, it appears to be permissible to include it in the database because of the well-established phenomenon of 'turnover' (e.g. MacArthur and Wilson 1967; Russell *et al.* 2006; McCollin 2015). The latter's impressive analysis of the 'extended dataset' from 1928 to 2002 for Skokholm Island, Wales, shows how informative a long-term record of breeding land birds can be and how the pattern can change over time. The Skokholm data might be described as close to the ideal as is possible for an offshore island (despite the 'hiatus' of the Second World War 1940–45 when no observations of breeding land birds were made) and it does not appear to have any equal in the British Isles at least (see McCollin 2015 and associated papers). Among the important findings of turnover is that an island's land bird avifauna is not unvarying, and species come and go and are often 'replaced' by different species. Therefore, the presence of a breeding land bird species at some time in the past is a valid record and therefore is a valid datum. Thus, despite the limitations in the dataset being treated here, we believe that some, albeit preliminary, generalisations can be made.

Bird species breeding on Irish offshore islands

A total of 57 species have been recorded as breeding on Irish islands (Fig. 4). However, some species occur on most if not all islands (rock pipit) while others including the long-tailed tit and siskin *Carduelis spinus* (Linnaeus) have only been found breeding on one island. Of the three endemic taxa in Ireland, i.e. the jay *Garrulus glandarius hibernicus* Witherby and Hartert, dipper *Cinclus cinclus hibernicus* Hartert and coal tit *Periparus ater hibernicus* (Ingram), only the dipper and coal tit have been recorded as breeding species on Irish islands. The dipper did breed on Clare Island (Ussher 1912) but had become extinct by 1945 (Barlee and Ruttledge 1945); there have been subsequent sightings (D'Arcy 1992 and John Feehan, pers. comm. November 2017; see Chapter 2) but there is no evidence of it having bred since Ussher's study. Finally, the coal tit has been found on only one island—namely Rathlin (Appendix 1).

The absence of the jay as a breeding species on Irish islands can presumably be explained by the fact it is a woodland specialist; interestingly it does not breed either on the Isle of Man (Sharpe 2007) or the Northern Isles of Scotland (Balmer *et al.* 2013).

The treecreeper *Certhia familiaris* Linnaeus, a specialist of deciduous and coniferous forests has never, so far as is known, bred on any Irish offshore island, but it has been recorded as a post-breeding season vagrant on Great Saltee, Cape Clear and the Copeland Islands off County Down (Hutchinson 1989). The treecreeper does breed on the Isle of Man, the Isle of Lewis and many of the Inner Hebrides but is absent from the Shetlands and the Orkneys (Sharpe 2007; Balmer *et al.* 2013).

The total absence of breeding bullfinches[1] *Pyrrhula pyrrhula* (Linnaeus) from all surveyed Irish offshore islands is less easily explained (it has bred on the inshore Achill Island; Hutchinson 1989). This sedentary species does not breed, or winter, on the Isle of Man (Sharpe 2007) and is also absent as a breeding species from 'many Scottish islands' (Balmer *et al* 2013) but is a relatively common sedentary breeding species of the Irish 'mainland'; it is a 'very rare autumn migrant' (Sharrock 1973) to, for example, Cape Clear (see also Hutchinson 1989; Gibbons *et al.* 1993; Balmer *et al.* 2013; Bayley and Caffrey 2018). This double-brooded species (see Ferguson-Lees *et al.* 2011) appears to have a broader ecological niche than the treecreeper, but like the latter it has been unable to colonise any Irish offshore island.

Perhaps the most remarkable absence of a species from the Irish island land bird 'community' is the rook. This species is both widespread (i.e. high *D*) and very numerous (i.e. high *N*) (the sixth most abundant species in Ireland; Fig. 1; Crowe *et al.* 2014), and it has the highest *M* or biomass of any land bird in Ireland (Fig. 2). Moreover, it is worth noting that rook flocks regularly visit Irish offshore islands including Cape Clear (Sharrock 1973) and Clare Island (Ussher 1912; see Chapter 2). The absence of the rook as a

[1] A pair reared two broods at Cape Clear (Cork) in 2016 (Bayley and Caffrey 2017).

breeding species may be related to its coloniality and wintering behaviour (see Chapter 8).

Of the 57 species found to be breeding on Irish islands, 15 (26.8%) are migrants; absent species include the reed warbler *Acrocephalus scirpaceus* (Hermann), common redstart *Phoenicurus phoenicurus* (Linnaeus), wood warbler *Phylloscopus sibilatrix* (Bechstein) and yellow wagtail *Motacilla flava* Linnaeus. All these species occur as scarce migrants on well-watched islands such as Great Saltee and Cape Clear, but none has bred at any of these.

The species area relationship

The species richness of Irish offshore islands is significantly correlated with the area of those islands in the sample (n = 29) (Fig. 5). The well-known and well-established species area relationship (MacArthur and Wilson 1967; Losos and Ricklefs 2010) is also confirmed in the NMDS analysis (Table 1, Fig. 6). This latter analysis also found that the presence of a lentic waterbody >100m^2, and the presence of human habitations either current or past has a statistically significant influence on the species richness of land birds on Irish offshore islands (Table 1). Latitude emerges as a correlate, but this may be due to the high proportion of the 29 islands in our sample that lie off the northwest and northern coastlines. Interestingly, longitude, or distance from the mainland, does not appear to exert an influence on island land bird species richness (Table 1). Nevertheless, it should be noted that distance from the 'mainland' in relation to Irish offshore islands shows relatively little variation; most islands lie within 10km of the nearest coastline. Consequently, with such little variation in this variable it is not surprising that 'distance to the nearest mainland' has not proved to be a significant determinant of island land bird species richness.

While the wheatear and rock pipit are among the most frequent breeding birds on Irish islands (Fig. 4) neither can be described as synanthropic, but the evacuation of islanders from the Great Blasket shows that other species are, including the wren, house sparrow *Passer domesticus* (Linnaeus) and song thrush *Turdus philomelos* Brehm (see Alexander 1954; Munns 1956; Hounsome and Rear 1966). Thus, presence of human habitations and associated agricultural and horticultural activities enable some land bird species to colonise offshore islands.

The Sørensen hierarchical clustering results, while interesting (Fig. 7), would obviously be strengthened if other avian groups were to be included, especially the raptors, owls, shorebirds and wildfowl. Their inclusion might produce a very different configuration of the dendrogram (see McCune and Grace 2002). Much remains to be investigated with this type of analysis.

Chough. Photo: R.T. Mills

Species pool size

In this study we aimed to expand the concept of 'pool size' by investigating, in separate regressions, the possible role of numerical abundance (*N*), biomass (*M*) and geographical distribution (*D*) on the frequency of different land bird species (*Y*) in our sample of 29 islands.

However, in a parallel study (Kelly *et al.* in prep.) we found that while on the island of Ireland *N* is highly correlated with *D*, bird weight is not. These, and other findings, led us to predict that the same interactions would obtain on the islands, i.e. that *N* would explain the frequency of occupancy (*Y*) and, by contrast, the weight of a bird species and its biomass (*M*) would not.

However, our findings were of the opposite sign; there is no correlation between *Y* and *N* but there is between *Y* and *M* and indeed, separately, the weight of individual bird species. Therefore, the configuration of the offshore island land bird assemblage appears to be noticeably different to that of the island of Ireland as a whole.

Firstly, while the trend between *N* and *Y* is positive, the slope, unlike that between *N* and *D* on the 'mainland', is not statistically significant. The reason appears to be that a high frequency of offshore islands are occupied by species that are coastal specialists and therefore, relatively speaking, occur in low numbers, e.g. the rock pipit is estimated to number 25,100 individuals and is present in 227 atlas squares, but has been recorded on 100% of our offshore island sample; wheatear (48,620 individuals and 328 atlas squares) occurs on 86% of offshore islands, and the chough, while found on 69% of the island sample, occupies only 116 atlas squares and numbers less than 2,000 individuals (Gibbons *et al.* 1993). In marked contrast, the Paridae species occur in numbers exceeding one million individuals and in more than 500 atlas squares, but on very few islands (Fig. 4; see also Gibbons *et al.* 1993).

Thus, it would appear that 'pool size' as defined purely by the number of individuals breeding on the island of Ireland is not a good predictor of the probability of occupancy of an offshore island.

Raven. Photo: R.T. Mills

Secondly, the prediction that biomass (*M*) and the weight of an individual bird species would have little or no influence on the frequency of occupancy on offshore islands was not met. On the contrary, biomass is significantly positively correlated with the frequency of occupancy of both the 'all island' and 'larger island' samples. This is probably explained by the relatively high frequency on offshore islands of some of Ireland's heaviest native terrestrial bird species such as the raven, (n = 71,500; 785g; 69% island occupancy), hooded crow (n = 537,830; 570g; 66% island occupancy and of course the chough (287.5g), which, as mentioned, also occurs on 69% of the islands.

The weight of individual bird species—and biomass—disparity between the offshore islands and the 'mainland' of Ireland does not appear to be an example of the 'Island Rule', also known as the 'Island Syndrome' (e.g. Clegg and Owens 2002; Lomolino 2005; Owens *et al.* 2006; Luther and Greenberg 2011). In fact, proof of this adaptive phenomenon was not directly addressed by us and would require a focused study of island bird morphology and biometrics such as that of Förschler and Geiter (2010) and Luther and Greenberg (2011), although, for example, in this context, the specimens collected during the original Clare Island survey (Ussher 1912) would be worth a detailed re-examination.

Interestingly, a previous study (Olson *et al.* 2009) found, in relation to global size variation in birds, that 'island assemblages have larger median body sizes than those of mainland assemblages at similar latitudes' but this asymmetry, as the authors' pointed out, may be due in part to the presence of seabirds on the islands in their sample (Olson *et al.* 2009).

Hooded crow. Photo: R.T. Mills

However, while the disparity in land bird weight and biomass of the Irish offshore islands may not be adaptive, at least in the sense of the 'Island Rule' concept, it is probably associated with the role of 'generalist foraging' (Scott *et al.* 2003), a feeding habit exemplified by most crow species (e.g. Coombs 1978; Ratcliffe 1997). This is particularly true of the raven and hooded crow, which are both abundant and widespread on, for example, the offshore islands of Scotland (Balmer *et al.* 2013). Both species have catholic diets, can forage on vegetable matter and are able to exploit the littoral at low tide (Ewins *et al.* 1986; Marquiss and Booth 1986; Ratcliffe 1997).

Similarly, in Ireland, the hooded crow, jackdaw, rook and raven (Berrow *et al.* 1991, 1992a; Berrow 1992; Smiddy 2016) have highly flexible opportunistic diets including littoral invertebrates. In addition, the hooded crow drops bivalves (mussels *Mytilus edulis*, cockles *Cerastoderma edule*) and periwinkles (*Littorina littorea*) to break them open and consume the contents (Davenport *et al.* 2014), and caches mussels (and probably other marine bivalves) for retrieval, when they gape, some days later (Berrow *et al.* 1992b).

This 'generalist foraging' in the hooded crow and many other corvid species enables it to meet its daily food intake demands of up to 33% of its body weight; see Yom-Tov (1974) in a study of the near identical sized carrion crow *Corvus corone* Linnaeus. Similar *pro rata* intake requirements are probably true of the jackdaw (present on nine; 31% of the Irish offshore islands) and the rook, which is absent as a breeding species from all offshore islands in Ireland (see Chapter 8). The latter example (rook) shows that a flexible diet, high body weight, high numerical abundance (>$2*10^6$) and a wide distribution (824 atlas squares, compared to only 380 in the raven) is not universally associated with the ability to colonise, and sustain a breeding population, on Irish offshore islands.

Nevertheless, the relatively high frequency of crows on offshore islands of Ireland together with the high level of persistence of their populations (Pimm *et al.* 1993) appears to explain the weight disparity with the 'mainland'.

Finally, while the *D* on *Y* regression was positive it was not statistically significant. However, as would be expected, several of the species with a high percentage of occupancy on the offshore islands are effectively confined to coastal atlas squares (e.g. rock pipit, rock dove, and chough; see Gibbons *et al.* 1993 and Balmer *et al.* 2013) and this means that the total number occupied is markedly smaller than most of the more terrestrial species, with a proportional effect on the slope of *D* on *Y*.

The avifauna of Ireland is 'on the edge' of the European land mass (Fuller *et al.* 2007; Kelly 2008) and many of the Atlantic offshore islands lie at or near the western extremity of the continent. This suggests that the birds of the Irish marine islands and indeed those off the coast of Great Britain would be worthy of much more detailed investigations than have been undertaken heretofore.

Ideally, however, a future study of the land bird avifauna of Irish offshore islands should involve dedicated visits in the same year to all the islands in the sample. Proof of breeding should be based, where possible, on the criteria of Balmer *et al.* (2013) and Ferguson-Lees *et al.* (2011). An exemplary study of this kind is that of Lysaght (2002), which also related the findings on the breeding birds of the Aran Islands to a simultaneous investigation of the local 'species pool'.

Acknowledgements

All the authors' fieldwork associated with this project was privately funded. We are indebted to Ms Rachel Hutchinson-Kelly for assistance with the production of this paper. Stuart Warner produced the map of the offshore islands included in this study. Finally, we would like to thank the referee for a very detailed but constructive review of the original draft; the critique has greatly improved the detail and accuracy of this manuscript.

REFERENCES

Abbott, I. 1980 Theories dealing with the ecology of landbirds on islands. *Advances in Ecological Research* **11**, 329–71.

Alexander, S.M.D. 1954 The birds of the Blasket Islands, with special reference to the Great Blasket Island, Inishvickillaun and Illaunboy and some notes on the adjacent mainland. *Bird Study* **1**, 148–68.

Avise, J.C. 1987 Mitochondrial DNA and the evolutionary genetics of higher animals. *Philosophical Transactions of the Royal Society London* **312B**, 325–42.

Avise, J.C. 1991 Ten unorthodox perspectives on evolution prompted by comparative population genetic findings on mitochondrial DNA. *Annual Review of Genetics* **25**, 45–69.

Avise, J.C. and Walker, D.E. 1998 Pleistocene phylogeographic effects on avian populations and the speciation process. *Proceedings of the Royal Society* **265B**, 457–63.

Balmer, D., Gillings, S., Caffrey, B., Swann, B., Downie, I. and Fuller, R. 2013 *Bird Atlas 2007–2011; the breeding and wintering birds of Britain and Ireland*. Thetford. BTO Books.

Barlee, J. and Ruttledge, R.F. 1945 Notes on the present status of birds on Clare Island. *Irish Naturalists' Journal* **8**, 311–3.

Barrington, R.M. 1900 *Migration of birds as observed at Irish lighthouses and lightships*. Dublin and London. Ponsonby and Porter.

Bayley, S. and Caffrey, B. 2017 *Cape Clear Bird Observatory Annual Report 2016*. Report Number 25. Wicklow. BirdWatch Ireland.

Bayley, S. and Caffrey, B. 2018 *Cape Clear Bird Observatory Annual Report 2017*. Report Number 26. Wicklow. BirdWatch Ireland.

Berrow, S. 1992 The diet of coastal breeding ravens in Co. Cork. *Irish Birds* **4**, 555–8.

Berrow, S., Kelly, T.C. and Myers, A.A. 1991 Crows on estuaries: distribution and feeding behaviour of the Corvidae on four estuaries in southwest Ireland. *Irish Birds* **4**, 393–412.

Berrow, S., Kelly, T.C. and Myers, A.A. 1992a The diet of coastal breeding hooded crows *Corvus corone cornix*. *Ecography* **15**, 337–46.

Berrow, S., Kelly, T.C. and Myers, A.A. 1992b The mussel caching behaviour of hooded crows *Corvus corone cornix*. *Bird Study* **39**, 115–9.

Blackburn, T.M., Brown, V.K., Doube, B.M., Greenwood, J.J.D., Lawton, J.H. and Stork, N.E. 1993 The relationship between abundance and body size in natural animal assemblages. *Journal of Animal Ecology* **62**, 519–28.

Blackburn, T.M., Gaston, K.J., Quinn, R.M., Arnold, H. and Gregory, R. 1997 Of mice and wrens: the relation between abundance and geographic range size in British mammals and birds. *Philosophical Transactions of the Royal Society of London* **352B**, 419–27.

Blackburn, T.M., Cassey, P. and Gaston, K.J. 2006 Variations on a theme: sources of heterogeneity in the form of the interspecific relationship between abundance and distribution. *Journal of Animal Ecology* **79**, 1426–39.

Boyle, D.P. 1997 A breeding survey of seabirds and other selected species on Cape Clear Island in 1995/1996. In Preston, K. (ed) *Cape Clear Bird Observatory. Report Number Twenty Four*. 37–44. Dublin. Irish Wildbird Conservancy.

Butlin, R.K., Galindo, J. and Grahame, J.W. 2008 Sympatric, parapatric or allopatric: the most important way to classify speciation. *Philosophical Transactions of the Royal Society* **363B**, 2997–3007.

BWP *Birds of the Western Palaearctic Interactive*. BWPi Version 2.

Claramunt, S., Derryberry, E.P., Remsen J.V. and Brumfield, R.T. 2012 High dispersal ability inhibits speciation in a continental radiation of passerine birds. *Proceedings of the Royal Society London* **279B**, 1567–74.

Clegg, S.M. and Owens, I.P.F. 2002 The 'island rule' in birds: medium body size and its ecological explanation. *Proceedings of the Royal Society, London.* **269B**, 1359–65.

Colhoun, K. and Cummins, S. 2013 Birds of conservation concern in Ireland. *Irish Birds* **9**, 523–44.

Collins, T.C. 1999 The Clare Island survey of 1909–1911: participants, papers and progress. In C. Mac Carthaigh and K. Whelan (eds) *New Survey of Clare Island* Volume 1, 1–40. Dublin. Royal Irish Academy.

Connor, E.F. and McCoy, E.D. 1979 The statistics and biology of the species area relationship. *American Naturalist* **13**, 791–833.

Coombs, F. 1978 *The Crows*. London. Batsford.

Crowe, O., Coombes, R.H., Lysaght, L., O'Brien, C., Choudhury, K.R., Walsh, A.J., Wilson, H.J. and O'Halloran, J. 2010 Population trends of widespread breeding birds in the Republic of Ireland 1998–2008. *Bird Study* **57**, 267–80.

Crowe, O., Musgrove, A. and O'Halloran, J. 2014 Generating population estimates for common and widespread breeding birds in Ireland. *Bird Study* **61**, 82–90.

D'Arcy, G. 1992 Fauna observed on Clare Island, Co. Mayo June 14–18 1990. *Irish Naturalists' Journal* **24**, 33–4.

Darwin, C. 1859 *On the origin of species by means of natural selection, or the preservation of favoured races in the struggle for life*. London. John Murray.

Davenport, J., O'Callaghan, M.J.A., Davenport, J.K. and Kelly, T.C. 2014 Mussel dropping by Carrion and hooded crows: biomechanical and energetic considerations. *Journal of Field Ornithology* **85**, 196–205.

Diamond, J.M. and May, R.M. 1977 Species turnover rates on islands: dependence on census interval. *Science* **197**, 266–70.

Dunning, J.B. (ed) 2008 *CRC Handbook of avian body masses*. Second edition. London. CRC Press.

Ewins, P.J., Dymond, J.N. and Marquiss, M. 1986 The distribution, breeding and diet of ravens *Corvus corax* on Shetland. *Bird Study* **33**, 110–6.

Ferguson-Lees, J., Castell, R. and Leech, D. 2011 *A field guide to monitoring nests*. Thetford. British Trust for Ornithology.

Förschler, M.I. and Geiter, O. 2010 Morphological shifts in house sparrows *Passer domesticus* in Helgoland. *Ornis Fennica* **87**, 93–8.

Fuller, R.J. (ed) 2012 *Birds and habitat. Relationships in changing landscapes.* Cambridge. Cambridge University Press.

Fuller, R.J., Gaston, K.J. and Quine, C.P. 2007 Living on the edge: British and Irish woodland birds in a European context. *Ibis* **149** 56–63.

Gaston, K.J. 1994 *Rarity*. London. Chapman and Hall.

Gaston, K.J., Blackburn, T.M., Greenwood, J.J.D., Gregory, R.D., Quinn, R.M. and Lawton, J.H. 2000 Abundance occupancy relationships. *Journal of Applied Ecology* **37** (Supplement 1), 39–59.

Gibbons, D.W., Reid, J.B. and Chapman, R.A. 1993 *The New Atlas of Breeding Birds in Britain and Ireland: 1988–1991*. London. Poyser.

Hillis, J.P. 2010 Rare Irish breeding birds, 2009. The annual report of the Irish Rare Breeding Birds Panel (IRBBP). *Irish Birds* **9**, 67–76.

Hounsome, M.V. and Rear, D. 1966 The bird population of the Great Blasket Island, Illaunboy and Beginish, Co. Kerry. *Irish Naturalists' Journal* **15**, 169–75.

Hutchinson, C.D. 1989 *Birds in Ireland*. Calton. Poyser.

Kelly, T.C. 2008 The origin of the avifauna of Ireland. In Davenport, J.L., Sleeman, D.P. and Woodman, P.C. (eds) *Mind the gap: postglacial colonization of Ireland*. 97–107. Special Supplement 2008. Belfast. Irish Naturalists' Journal.

Lawton, J.H. 1989 What is the relationship between population density and size in animals? *Oikos* **55**, 429–34.

Lloyd, C.S. 1984 The birds of Clare Island, Co. Mayo in June 1982. *Irish Naturalists' Journal.* **21**, 212–6.

Lomolino, M.V. 2005 Body size evolution in insular vertebrates: generality of the island rule. *Journal of Biogeography* **32**, 1683–99.

Losos, J.B. and Ricklefs, R. (eds) 2010 *The theory of island biogeography revisited*. Princeton and Oxford. Princeton University Press.

Luther, D. and Greenberg, R. 2011 The island syndrome in coastal wetland ecosystems: convergent evolution in large bills in mangrove passerines. *The Auk* **28**, 201–4.

Lysaght, L. 2002 *An Atlas of Breeding Birds of the Burren and the Aran Islands*. Dublin. BirdWatch Ireland.

MacArthur, R.H. and Wilson, E.O. 1963 An equilibrium theory of insular zoogeography. *Evolution* **17**, 373–87.

MacArthur, R.H. and Wilson, E.O. 1967 *The theory of island biogeography*. Monographs in Population Biology Volume 1. Princeton. Princeton University Press.

Marquet, P.A., Navarette, S.A. and Castilla, J.C. 1995 Body size, population density, and the Energetic Equivalence Rule. *Journal of Animal Ecology* **64**, 325–32.

Marquiss, M. and Booth, C.J. 1986 The diet of ravens in Orkney. *Bird Study* **33**, 190–5.

McCollin, D. 2014 Abundance estimates for landbirds and seabirds extracted and compiled from annual reports of the Skokholm bird observatory. http://doi.pangaea.de/10.1594/ PANGAEA 833759.

McCollin, D. 2015 The curious case of Skokholm: equilibrium, non-equilibrium and phase shift in an island bird assemblage. *Ecography* **38**, 986–91.

McCune, B. and Grace, J.B. 2002 *Analysis of Ecological Communities*. Oregon. MjM Software Design.

McGuinness, K.A. 1984 Equations and explanations in the study of species-area curves. *Biological Reviews* **59**, 423–40.

McKechnie, A.F. 2015 Global variation in avian metabolic rates and the slow pace of life of tropical birds. *Functional Ecology* **29**, 297–8.

Munns, D.J. 1956 Further notes on the birds of the Blasket Islands. *Bird Study* **3**, 248–50.

Nairn, R. and O'Halloran, J. (eds) 2012 *Bird Habitats in Ireland*. Cork. Collins Press.

Oksanen, J., Blanchet, F.F, Kindt, R., Legendre, P. and Minchin, P.R. 2016 Vegan Community Ecology. R package version.

Olson, V.A., Davies, R.G., Orme, C.P.L., Thomas, G.H., Meiri, S., Blackburn, T.M., Gaston, K.J., Owens, I.P.F. and Bennett, P.M. 2009 Global biogeography and ecology of body size in birds. *Ecology Letters* **12**, 249–59.

O'Regan, S.M., Flynn, D., Kelly, T.C., O'Callagahan, M.J.A., Pokrovskii, A. and Rachinskii, D. 2012 The response of the Woodpigeon (*Columba palumbus*) to relaxation of intraspecific competition: a hybrid modelling approach *Ecological Modelling* **224**, 54–64.

Owens, I.P.F., Scott, S.N., Robinson, S.I., Clegg, S.M. and Kikkawa, J. 2006 Large body size in island-dwelling passerines: the roles of insular specialisation, niche expansion and ecological release. *Acta Zoologica Sinica* **52** (Supplement) 262–6.

Passy, S.J. 2012 A hierarchical theory of macroecology. *Ecology Letters* **15**, 923–4.

Pimm, S.L., Diamond, J., Reed, T.M., Russell, G.J. and Verner, J. 1993 Times to extinction for small populations of large birds. *Proceedings of the National Academy of Sciences, USA* **90**, 10871–5.

Ratcliffe, D. 1997 *The Raven*. London. Poyser.

Reed, T. 1980 Turnover frequency in island birds. *Journal of Biogeography* **7**, 329–35.

Reed, T. 1981 The number of breeding landbird species on British islands. *Journal of Animal Ecology* **50**, 613–24.

Reuman, D.C., Mulder, C., Benasek-Richter, C., Blandanier, M-F.C., Breure, A.M., Hollander, H.D., Knietal, J.M., Rafaelli, D., Woodward, G. and Cohen, J.E. 2009 Allometry of body size and abundance in 166 food webs. *Advances in Ecological Research* **41**, 1–44.

Ridley, M. 2004 *Evolution*. Third edition. Oxford. Blackwell.

Rosenzweig, M.L. and Ziv, Y. 1999 The echo pattern of species diversity: pattern and process. *Ecography* **22**, 614–28.

Russell, G.J., Diamond, J.M., Pimm, S.L. and Reed, T.M. 1995 A century of turnover: community dynamics on three timescales. *Journal of Animal Ecology* **64**, 628–41.

Russell, G.J., Diamond, J.M., Reed, T.M. and Pimm, S.L. 2006 Breeding birds on small islands: island biography or optimal foraging? *Journal of Animal Ecology* **75**, 324–39.

Scott, S.N., Clegg, S.M., Blomberg, S.P., Kikkawa, J. and Owens, I.P.F. 2003 Morphological shifts in island-dwelling birds: the roles of generalist foraging and niche expansion. *Evolution* **57**, 2147–56.

Sharpe, C.M. 2007 *Manx Bird Atlas*. Liverpool. Liverpool University Press.

Sharrock, J.T.R. 1973 *The Natural History of Cape Clear Island*. Berkhamsted. Poyser.

Simberloff, D. 1976 Experimental zoogeography of islands: effects of island size. *Ecology* **57**, 629–48.

Smiddy, P. 2010 A systematic list of Irish Birds. *Irish Birds* **9**, 130–4.

Smiddy, P. 2016 Diet of coastal breeding ravens *Corvus corax* in east County Cork. *Irish Birds* **10**, 335–8.

Storch, D. and Gaston, K.J. 2004 Untangling ecological complexity on different scales of space and time. *Basic and Applied Ecology* **5**, 389–400.

Stracey, C.M. and Pimm, S.L. 2009 Testing island biogeography theory with visitation rates of birds to British islands. *Journal of Biogeography* **36**, 1532–9.

Sutherland, W.J. *et al.* 2013 Identification of 100 fundamental ecological questions. *Journal of Ecology* **161**, 58–67.

Ussher, R.J. 1912 A biological survey of Clare Island in the county of Mayo, Ireland and adjoining district. *Proceedings of the Royal Irish Academy* **31**, Sec.11, Part 20.

Whittaker, R.J. and Fernandez-Palacios, J.M. 2007 *Island Biogeography*. Second edition. Oxford. Oxford University Press.

Whittaker, R.J., Triantis, K.A. and Ladle, R.J. 2008 A general dynamic theory of oceanic island biogeography. *Journal of Biogeography* **36**, 977–94.

Williams, J.B., Miller, R.A., Harper, J.M. and Wiersma, P. 2010 Functional linkages for the pace of life, life-history and environment in birds. *Integrative and Comparative Biology* **50**, 855–68.

Woodward, F.I., Quaife, T. and Lomas, M.R. 2010 Changing climate and the Irish landscape. *Biology and Environment: Proceedings of the Royal Irish Academy* **110B**, 1–16.

Yom-Tov, Y. 1974 The effect of food and predation on breeding density and success, clutch size, and laying date of the crow (*Corvus corone* L.). *Journal of Animal Ecology*, **43**, 479–98.

CHAPTER 7

THE EQUILIBRIUM NUMBER OF LAND BIRD SPECIES ON CLARE ISLAND

T.C. Kelly, P. Smiddy, E. McGreal, P. Winters, U. Keating and R.E. Cussen

ABSTRACT

This study had two aims: (i) to estimate the equilibrium number of breeding land bird species on Clare Island and (ii) to describe the apparent turnover of species *T* since the findings of the first survey (1909–11). Although 35 land bird species are known to have bred at least once over the 104-year interval, the equilibrium number of species (i.e. where the curve of extinctions and immigrations intersect) is estimated to be 24. This, interestingly, is four less than the total number of species (n = 28) listed by Ussher (1912). There have been seven extinctions (*E*) and six immigrations (*I*) since the first survey. Extinct species include the corn bunting *Emberiza calandra* (Linnaeus), dipper *Cinclus cinclus* (Linnaeus), common whitethroat *Sylvia communis* Latham, sand martin *Riparia riparia* (Linnaeus), common cuckoo *Cuculus canorus* Linnaeus and twite *Carduelis flavirostris* (Linnaeus). Immigrants include the willow warbler *Phylloscopus trochilus* (Linnaeus), tree sparrow *Passer montanus* (Linnaeus) and chaffinch *Fringilla coelebs* Linnaeus. The rate of apparent turnover *T* was estimated by employing the equation of Diamond and May (1977). The analysis confirmed that the longer the interval between successive surveys the lower the rate of 'apparent turnovers' *T*. However, because of the probable influence of a variety of factors—especially the very long gaps between some of the successive surveys—the results should be regarded as preliminary rather than definitive.

Introduction

The theory of island biogeography of MacArthur and Wilson (1963, 1967) proposes that there is a dynamic equilibrium in the species richness of an island depending on the 'balance' of the opposing processes of immigration and extinction (Simberloff 1974). 'Accordingly, the biota of an island reaches a steady state at which the appearance of a new species equals the disappearance of the residents' (Ricklefs and Bermingham 2001). This 'steady state' is more usually termed the equilibrium number of species and will be found where the curve of increasing extinctions intersects with that of decreasing immigrations.

However, while the equilibrium number will remain, broadly speaking, the same over time, the species composition will change; this is known as 'turnover' (e.g. Simberloff 1974; Hutchinson 1989). Turnover, and the equilibrium number, will vary primarily with the areas of islands (the species-area relationship (SAR); and the island species-area relationship (ISAR) e.g. Triantis *et al.* 2012) and their distances from the mainland, i.e. the sources of potential propagules (MacArthur and Wilson 1967; Losos and Ricklefs 2010; McCollin 2015, 2017).

Neither the equilibrium number, nor the turnover in species composition can be easily quantified with exactitude, especially, as is often the case, when the surveys of island breeding land birds are intermittent, with intervals of several years or decades between consecutive expeditions (Diamond and May 1977). For example, few islands around the coast of Britain and Ireland have a long-term record of continuous land bird breeding season monitoring as Skokholm Island in Wales, which, apart from the war years of 1940–45, has been surveyed annually since 1928 (McCollin 2015, 2017). The accumulated data has been thoroughly analysed (e.g. Williamson 1981, 1989) though in certain aspects reaching different conclusions (e.g. McCollin 2015, 2017). For example, Williamson (1983) states that 'most of the turnover treated by MacArthur-Wilson theory is ecologically trivial', whereas others, e.g. Heaney (2000) and McCollin (2015, 2017) do not appear to support this conclusion. Turnover, or even disequilibrium, is of both ornithological and ecological interest. For example, gains, or losses, in species diversity are widely believed to influence the productivity, stability and resilience of ecosystems in which they reside (e.g. Tilman 1996; McCann 2000; Pennekamp *et al.* 2018). Apart from any other considerations, the fortunes of land bird populations on islands may reflect in part how the species is performing on the 'mainland' (McCollin 2015, 2017). Therefore, the permanent extinction (aside from that caused by demographic accident e.g. Pimm *et al.* 1988; Pimm and Raven 2000) or colonisation, of an island by a land bird species may be of considerable ornithological interest.

The turnover of land birds on islands off the coast of Ireland (or elements related to it, i.e. time to extinction) have been extensively studied (Reed 1980, 1981; Pimm *et al.* 1993; Russell *et al.* 1995, 2006; Manne *et al.* 1998; Stracey and Pimm 2009); most of this research has focused on Cape Clear and the Copeland Islands, where, unlike Clare Island, there are bird observatories. However, apart from Lloyd (1984), this aspect of the ecology of Clare Island does not appear to have been investigated in detail before now.

Therefore, the aims of this study were (i) to estimate the equilibrium number of land bird species on Clare Island and (ii) to give an account of the apparent 'turnover' in species richness that has occurred since the first survey of 1909–11 (Ussher 1912).

Methods

Preamble

One of the most intractable problems in estimating the apparent rate of 'turnover' T is the interval of time t between successive censuses—a problem discovered and elegantly described by Diamond and May (1977). This gives rise to an 'uncertainty principle' that can lead to considerable underestimates of the apparent rate of turnover; 'the longer t is the more cases of turnover (either immigrations or extinctions) will be observed. However, with increasing t, T will underestimate the true turnover rate by an increasing factor because a species may repeatedly become extinct and then an immigrant (or vice versa) in the time between censuses (the so called in–and–out effect)' (Diamond and May 1977).

The above quoted sentences from Diamond and May (1977) puts in context the difficulties of estimating T and the equilibrium number of land bird species on Clare Island. For instance, it is a remarkable fact that there appears to have been no survey of the land birds of Clare Island between 1912 and that of Barlee and Ruttledge (1945)—an interval of 36 years since the first survey that spanned the years 1909–11 (Ussher 1912). In fact, the minimal interval is 34 years, and the maximum between the surveys is 36 years, thus adding to the complication of compiling the estimate of T (see Table 1). A further 27 years elapsed before the next census, that of the first *Atlas of breeding birds* (Sharrock 1976), which was conducted over the 1968–72 interval though again there is uncertainty over the actual length of the interval so a maximum and minimum number of years (t max and t min) had to be assigned (Table 2).

Data selection

A total of ten surveys have been included in the analysis spanning the 1909 to 2013 interval (Table 1). We applied some basic ground rules in selecting data for analysis. For example, we omitted records of a species (though, of course, including them as sightings) where there was inconclusive evidence of breeding, according to criteria specified in, for example, Table 2.2 in Balmer *et al.* (2013). We applied these 'rules' rigorously to reports of a species that had been absent from the island for many years, e.g. dipper *Cinclus cinclus* (Linnaeus) (D'Arcy 1992) and twite *Carduelis flavirostris* (Linnaeus) (Winters 2006). In addition, if the species was seen for the first time on Clare Island, proof of breeding had to be available. If this was not the case the species was not included in the analysis of breeding species though, again, the sighting is recorded (Table 1).

Table 1

Species of land bird believed to be breeding on Clare Island during some or all ten surveys (1912 to 2013) included in this analysis.

Breeding species	Ussher 1912	Barlee and Ruttledge 1945	Sharrock 1976	Lloyd 1984	Gibbons *et al.* 1993	Forsyth and Allen 1997	Coombes 2000 and Kelly 2000	Winters 2006	Balmer *et al.* 2013	Kelly 2013
Rock Dove	1	1	1	1	1	-	1	1	1	1
Woodpigeon	0	0	0	S	0	-	-	0	0	-
Collared Dove	0	0	0	0	0	S	-	0	0	S
Cuckoo	1	1	1	1	1	0	-	-	0	-
Swift	0	0	0	S	0	-	-	-	0	-
Skylark	1	1	1	1	1	1	1	1	1	1
Sand Martin	0	0	1	0	1	-	-	-	0	-
Swallow	1	0	1	1	1	1	1	1	1	1
House Martin	0	0	0	S	0	-	-	-	0	-
Meadow Pipit	1	1	1	1	1	1	1	1	1	1
Rock Pipit	1	1	1	1	1	1	1	1	1	1
Grey Wagtail	1	0	0	0	0	1	1	0	1	0
Pied Wagtail	1	1	1	1	1	1	1	1	1	1
Dipper	1	0	0	0	0	0	-	-	0	-
Wren	1	1	1	1	1	1	1	1	1	1
Dunnock	1	1	1	0	0	1	1	1	1	1
Robin	1	1	1	0	1	1	1	1	1	1
Stonechat	1	0	1	1	1	1	1	1	1	1
Wheatear	1	1	1	1	1	1	1	1	1	1
Blackbird	1	1	1	1	1	1	1	1	1	1
Song Thrush	1	0	1	0	1	0	1	1	1	1
Grasshopper Warbler	0	0	0	0	S	-	-	-	0	-
Sedge Warbler	1	0	0	0	1	1	1	1	1	1
Whitethroat	1	0	1	0	0	-	-	0	0	-
Willow Warbler	0	0	0	0	0	1	1	1	1	1
Goldcrest	0	0	0	0	0	-	-	1	0	-
Magpie	0	0	0	1	0	1	1	1	1	1
Chough	1	1	1	1	1	1	1	1	1	1
Jackdaw	0	0	0	0	0	S	-	-	0	-
Rook	0	S	0	0	0	S	-	-	0	-
Hooded Crow	1	1	1	1	1	1	1	1	1	1
Raven	1	1	1	1	1	1	1	1	1	1
Starling	1	0	1	0	1	1	1	1	1	1
House Sparrow	1	1	0	1	1	1	-	1	1	1
Tree Sparrow	0	0	1	0	1	1	1	1	1	1
Chaffinch	0	0	0	1	1	0	-	0	0	1
Goldfinch	0	0	0	0	0	-	-	S	S	S
Linnet	1	1	1	1	1	1	1	1	1	1
Twite	1	1	1	0	1	1	-	0	0	-
Lesser Redpoll	0	0	0	0	0	-	-	0	0	1
Yellowhammer	1	1	0	0	0	-	-	0	0	-
Reed Bunting	1	1	1	1	1	1	1	1	1	1
Corn Bunting	1	0	0	0	0	0	0	-	0	-
	28	**19**	**24**	**19**	**25**	**24**	**24**	**25**	**25**	**26**

Notes: 1 = Breeding; 0 = Absent; S = Seen but breeding not proved.

Table 2
List of surveys (n = 10) included in this analysis; the interval in years *t* between successive surveys is shown but with a max *t* and a min *t* assigned to incorporate the probable maximum and minimum number of years between censuses. The total turnover as well as the *T* the estimated rate of turnover is also given.

Surveys included for analysis	*t* interval length (Years)	Max *t* (Years)	Min *t* (Years)	N. species of land birds	N. E	N. I	Total T	*T* (Max *t*)	*T* (Min *t*)
Ussher 1909–11				28					
Barlee and Ruttledge 1945	36	36	34	19	9	0	9	0.0089	0.00945
1968–72 Atlas (Sharrock 1976)	27	27	23	24	2	9	11	0.019	0.023
Lloyd May and June 1982	14	14	10	19	7	3	10	0.04135	0.05789
1988–91 Atlas (Gibbons *et al.* 1993)	9	9	6	25	1	7	8	0.0355	0.0533
Forsyth 1996 and 1997	9	9	8	24	3	3	6	0.0231	0.026
Coombes and Kelly April-July 2000	4	4	3	24	1	1	2	0.021	0.028
Winters 2006	6	6	6	25	1	1	2	0.02	0.02
2007–11 Atlas (Balmer *et al.* 2013)	5	5	1	25	1	1	2	0.016	0.08
Kelly 2013	6	6	2	26	1	2	3	0.019	0.0577

Time intervals

As can be seen in Table 1 some of the surveys of Clare Island spanned several years, including the first census (Ussher 1912) as well as all the atlas investigations. Consequently, it is not possible to provide exact estimates of the number of years between successive surveys. Therefore, in calculating *T*, it is necessary to apply a maximum and minimum interval value for *t* (Table 1). We pooled all data collected during the first survey, and, similarly, with all the atlas studies and the field work of Forsyth and Allen (1997). In addition, R.H. Coombes surveyed the island in April 2000 as did T.C. Kelly who also undertook a census in July 2000. All these data have been combined to make one list for the year 2000 (Table 1).

Calculating T

Diamond and May (1977) developed an equation (eq. 1) to calculate the apparent 'turnover' rate *T* within the time interval *t*.

$$T(t) = (I + E)/(S_1 + S_2)t \text{ Eq. 1.}$$

We followed the narrative form given by Russell *et al.* (1995), i.e. the apparent turnover rate *T* = number of extinctions + numbers of immigrations/total number of species present x the census interval.

While additional 'turnover-related' mathematics has been developed (e.g. Russell *et al.* 1995, 2006), the present study is confined to the simplest form of analysis, because the data from Clare Island are rather limited and, additionally, burdened with uncertainty as to the exact value of *t*.

Results

Equilibrium number

Over the 104 years (1909 to 2013) of albeit haphazard surveys on Clare Island, a total of 35 breeding land bird species has been recorded (Table 1). It is estimated that *E*—the number of permanent extinctions—is seven. The number of apparently permanent immigrations *I* is six, so broadly speaking the land bird fauna is at equilibrium (Tables 2, 3). Interestingly, the highest number of species was recorded during the first survey (n = 28, Table 1), and it appears that something of a phase shift, *sensu* McCollin (2015), occurred thereafter. The mean number of species recorded following the first survey was 23.1 (±0.841, SE); the Median is 24 ± 1.9386, CI (95%) (Table 2). If the first survey is included the mean over the 104 years is 23.9 species (±0.9; Median 24.5 ± 2.036, CI (95%) (Table 2).

Therefore, the equilibrium number of species over the 104-year interval appears to be 24.

Turnover

In terms of apparent turnover, the data clearly shows that the longer the interval between successive surveys, the lower the estimate of *T* (mean *t* max = 0.027 ± 0.003, SE; mean *t* min 0.039 ± 0.0078, SE) (see also Table 2), thus confirming the clear-cut trend identified by Diamond and May (1977). No immigrant species were discovered in Barlee and Ruttledge's (1945) survey—the first since Ussher (1912) (Tables 2, 3)—but none of the absent species, i.e. dipper, corn bunting *Emberiza calandra* (Linnaeus) and common whitethroat *Sylvia communis* Latham were recorded as present and definitely breeding in any subsequent survey. Overall, however, seven species appeared to have become extinct on Clare Island over the 104-year interval since the first survey, namely the corn bunting, yellowhammer *Emberiza citrinella* Linnaeus, dipper, common cuckoo *Cuculus conorus* Linnaeus, sand martin *Riparia riparia* (Linnaeus), common whitethroat and twite (Table 3). Subsequent surveys showed six definite immigrations (willow warbler *Phylloscopus trochilus* (Linnaeus), sand martin, lesser redpoll *Carduelis cabaret* (P.L.S. Müller), chaffinch *Fringilla coelebs* Linnaeus, tree sparrow *Passer montanus* (Linnaeus) and goldcrest *Regulus regulus* (Linnaeus), approximately balancing the extinctions (Table 2, 3). Of course, some species appear to have recolonised the island for short intervals, including the sand martin and, possibly, the goldcrest but the chaffinch, which was found by Lloyd in 1982 (Lloyd 1984) was not found again until 2013 when it appears to have been widely distributed.

Other species were found to be breeding, e.g. the sparrowhawk *Accipiter nisus* (Linnaeus) raised four chicks in 1999 (observed by R.E. Cussen), but this raptor was not detected (though seen by D'Arcy in June 1989, but not in July 1990) during any of the subsequent dedicated surveys and the species does not appear to have bred since. D'Arcy (1992) recorded lesser redpoll in June 1989 but not July 1990; R.E. Cussen also found lesser redpolls on the island during the summer months of 1999 but these were not recorded as definitely breeding until 2013 and the species had not been encountered during the intervening surveys (e.g. Balmer *et al.* 2013). D'arcy (1992) heard a grasshopper warbler *Locustella naevia* (Boddaert) in June 1989 and again in July 1990.

Finally, three species show evidence of the 'in-and-out-effect' (Diamond and May 1977). The grey wagtail *Motacilla cinerea* Tunstall is perhaps the most obvious example, but the song thrush *Turdus philomelos* C.L. Brehm—except for the most recent surveys—also appears to show evidence of extinction followed by successful recolonisation; and the sand martin bred on Clare Island in the summer of 2019 (see Chapter 2).

Discussion

The total number of land birds that are known to have bred on Clare Island is 37 (i.e. inclusive of known extinctions before Ussher's 1912 survey),

Table 3
Species which became extinct and immigrated since the first survey, 1909–1911 (Ussher 1912).

Survey	Year(s)	E	I	Extinct species	Immigrant species
Ussher (1912)	1909–1911	-	-	-	-
Barlee and Ruttledge (1945)	1945	9	0	Corn Bunting, Starling, Whitethroat, Song Thrush, Stonechat, Sedge Warbler, Dipper, Grey Wagtail, Swallow	0
Sharrock (1976)	1968–72	2	9	Yellowhammer, House Sparrow	Swallow, Sand Martin, Stonechat, Song Thrush, Whitethroat, Starling, Tree Sparrow
Lloyd (1984)	1982	7	3	Sand Martin, Dunnock, Robin, Song Thrush, Whitethroat, Starling, Tree Sparrow	Magpie, House Sparrow, Chaffinch
Gibbons *et al.* (1993)	1988–91	1	7	Magpie	Robin, Sedge Warbler, Song Thrush, Starling, Tree Sparrow, Twite
Forsyth and Allen (1997)	1996–97	3	3	Song Thrush, Chaffinch, Cuckoo	Willow Warbler, Magpie, Dunnock
Coombes and Kelly (2000)	2000	1	1	Twite	Grey Wagtail
Winters (2006)	2006	1	1	Grey Wagtail	Goldcrest
Balmer *et al.* (2013)	2007–11	1	1	Goldcrest	Grey Wagtail
Kelly (2013)	2013	1	2	Grey Wagtail	Lesser Redpoll, Chaffinch

Emerging woodland at Ballytoohy More. This appears to be the main breeding site of the lesser redpoll *Carduelis cabaret* (P.L.S. Müller). Neither this wood nor that at Portlea have been comprehensively surveyed. Autumn surveys might prove very informative as to the number and variety of migrant species passing through Clare Island. Photo: T. Kelly

though possible turnover in small species, i.e. smaller and less conspicuous than larger easily detectable taxa, over the same interval, is totally undescribed. Therefore, the figure for the total number of species must be treated with caution as an even larger number of taxa may have been present in the 'turnovers' in the 1880s before the commencement of Ussher's (1912) study.

The maximum number of land bird species detected in the ten surveys reviewed here is 28 (Ussher 1912) and the minimum is 19 (Table 1) (Barlee and Ruttledge 1945; Lloyd 1984). The assumption, however, that the lower number of species recorded in these latter surveys is due to extinction, must be treated with caution. Several factors may militate against the detection of a variety of bird species during a survey of a complex insular landscape such as exists on Clare Island. These variables include the weather *during* and *before* the survey, the timing of the expedition, the duration of the field assessments and the total area of the island that has been investigated.

Weather is perhaps the key variable in determining the success (in terms of accurate and comprehensive land bird species detections) or otherwise, of offshore island surveys. However, travel arrangements will have to be made months, or at least weeks, in advance. If the survey area is large, then investigations will have extended over several days. Observers and surveyors will have to rely on as accurate weather forecasts as are available to ensure that optimal ambient conditions will obtain for the duration of the field investigations (e.g. Bibby *et al.* 2000). It is interesting to note, however, that weather forecasting skill has improved steadily over the past 40 years: 'for example forecast skill in the 3–10 days range has been increasing by about one day per decade: todays 6-day forecast is about as accurate as the 5-day forecast 10 years ago' (Bauer *et al.* 2015). It is therefore a fact, particularly in the 1980s and 1990s, that ideal conditions for field surveys would have existed more by luck than design.

For example, Lloyd (1984) stated that during her three-day survey from 21 to 24 June, 'the weather was wet and very windy' and this could have explained the absence of small birds like the robin *Erithacus rubecula* (Linnaeus), whitethroat and dunnock *Prunella modularis* (Linnaeus). Wet and very windy weather, as pointed out by Lloyd (1984) is not optimal for surveying small passerines breeding in hedgerows and thickets (see also Bibby *et al.* 2000).

Likewise, Forsyth and Allen (1997) achieved a full day's censusing on 8 May 1996, but an evening survey had to be 'abandoned because of heavy persistent rain'. They go on to state that 'the plan to census for another two days was also abandoned because of atrocious weather conditions', but good weather for surveying existed during the 18 to 20 July 1997 expedition (Forsyth and Allen 1997).

By contrast, Barlee and Ruttledge (1945) surveyed 'under excellent conditions' from 25 to 27 April 1945. It is clear from their remarks that the weather was ideal for both aural and visuals registrations. Thus, they ascribe the absence of the corn bunting, stonechat *Saxicola torquatus* (Linnaeus) and song thrush to the exceptionally severe weather in January of that year (see also Ruttledge 1950, 1994). However, Barlee and Ruttledge (1945) also state that the corn bunting and stonechat were present in 'near normal numbers' on the mainland coastline, so perhaps 'the weather cannot have been held entirely responsible'. Nevertheless, the total absence of the stonechat and song thrush is almost certainly indicative of an effect of the weather in the previous January (i.e. before the survey).

The timing of the Barlee and Ruttledge (1945) survey in late April 1945 may have been too early to detect both the sedge warbler *Acrocephalus schoenobaenus* (Linnaeus) and the whitethroat, both of which are migrant species whose arrival time is typically early to mid-May (see also Ruttledge 1950, 1994; Forsyth and Allen 1997). The timing of visits to Clare Island (excepting that of Winters (2006)) has rarely been fully optimal for the detection of both resident and migrant breeding land bird species (see Chapter 5). Ideally, a comprehensive breeding land bird survey of the whole island would require visits in mid to late April, May and June of the same year. However, exceptionally cold and wet springs—such as occurred in 2013—would have, undoubtedly, adversely affected surveys in April and May (see Chapter 5).

Hounsome and Rear (1966) remark that 'Great Blasket Island (459ha) is notoriously difficult to cover completely' and where Clare Island (1,594ha) is concerned a field trip of at least several days duration is required to survey all the habitats and all the different bird groups within its complex landscape. The 2013 survey walked approximately 16km of transects (32km there and back to base) over the 16 to 19 June interval in excellent viewing conditions. However, there was not time to visit the extensive areas of coastal habitat between the western end of Strake and the Signal Tower. Moreover, the 2013 survey did not attempt to census the seabird populations and was, in effect, exclusively focused on the land birds. Finally, because of the transect methods being used some emerging woodland habitat near Portlea was not surveyed.

Clearly, some of the smaller islands, e.g. Copeland (32ha), Great Saltee (89ha) and, in Wales, Skokholm (97ha) can be surveyed reasonably quickly, but much more time would have to be allocated to investigations on islands like Inishmore (3,090ha), Arranmore (1,754ha) and Rathlin (1,437ha). Intermittent surveys of the larger islands might not always find a very small founder territory and, equally, a tiny relict population that had survived a recent period of severe winter weather.

Either way, all the above-mentioned factors must be considered when attempting to estimate both the equilibrium number and the apparent 'turnover' T on an island. It is probably not an exact science and in the absence of the rigorous implementation of a standardised survey methodology the old saw that the 'absence of evidence is not evidence of absence' may apply where larger islands are surveyed infrequently.

Two species affected by the 'in-and-out-effect' appear to occur in the record of land bird surveys on Clare Island (see Tables 1 and 3), namely the grey wagtail and the song thrush. The grey wagtail may be influenced by hard winters to which they are very susceptible (Balmer *et al.* 2013) and, possibly, low rainfall in spring leading to reduced water levels in streams and drainage ditches (T.C. Kelly, personal observations). The grey wagtail has only been recorded on four (40%) of the ten surveys reviewed here (Table 1). The song thrush, which has been recorded on 70% of the Clare Island surveys reviewed here, is also susceptible to hard winters (Thompson *et al.* 1997; Balmer *et al.* 2013; McGeehan 2018) and this almost certainly explains its absence in the Barlee and Ruttledge (1945) survey. However, the song thrush is also adversely affected by predators, especially during the nestling stage of the breeding cycle (Kelleher and O'Halloran 2006). McGeehan (2018) cites an interesting example of the possible impact of one predator (magpie *Pica pica* (Linnaeus)) on the song thrush on Inishbofin where, in 2011, following the removal of 'more than a dozen pairs of magpies, the island's song thrush population soared from three pairs to more than fifteen pairs most of which are double brooded'.

Two land bird species had become extinct before Ussher's (1912) survey, namely red grouse *Lagopus lagopus* (Linnaeus) and golden eagle *Aquila chrysaetos* (Linnaeus); the last golden eagle appears to have been 'obtained' in October 1887; the specimen is in the Natural History Museum (Dublin). According to Ussher (1912) the golden eagle 'used to prey upon the grouse then plentiful on Clare Island'; this appears to have been a time when there was an 'absence of guns' on the island, though it is not known exactly when red grouse became extinct (Ussher 1912; Lloyd 1984; Ruttledge 1994). However, it is the removal of the *Calluna* formation—due to increased sheep grazing—that probably led to the extinction of the red grouse (Doyle and Foss 1986; see Chapter 2 and references therein).

Other groups

The equilibrium number of land bird species has been estimated to be 24. However, we did not include the rails (Rallidae), waders (Charadriidae and Scolopacidae), raptors (Accipitridae and Falconidae), wildfowl (Anatidae), seabirds (many families) and introduced species, i.e. pheasant *Phasanius colchicus* Linnaeus and Canada goose *Branta canadensis* (Linnaeus), in Table 4, from

Table 4
Total number of breeding bird species on Clare Island from Ussher's (1912) survey to the present. Note the extinctions before Ussher (1912) are omitted. Note: the situation regarding the rails remains uncertain, all three species were present in the first survey but only the moorhen *Gallinula chloropus* (Linnaeus) now occurs—though very infrequently—and mostly without definitive proof of breeding.

Avian group	Number of species	Extinctions	Immigrations	In equilibrium	Dis-equilibrium
Land birds	35	7	6	1	0
Raptors	2	0	0	1	0
Wading birds	5	0	0	1	0
Wildfowl	1	0	0	1	0
Rails	3	3	0	0	1
Seabirds	14	1	6	0	1
Introduced species	2	-	-	-	-
Totals	60	11	12	4	2

which it can be seen that a total of 60 bird species have bred at one time or the other, inclusive of those discovered in the first survey (Ussher 1912). Two species are known to have become extinct before the 1909–11 survey and two species have been introduced making a possible maximum total of 64 species. However, because of the long intervals between successive surveys the 'uncertainty principle' of Diamond and May (1977) must be considered and the period before the first survey, i.e. 1888 to 1909 has no data other than that relating to large and obvious species as explained above.

Nevertheless, it is interesting to note that when the non-land bird groups are included it appears that overall the breeding bird fauna is in equilibrium, i.e. $E = 11$ and $I = 12$ (Table 4). Equally, however, two groups are not in equilibrium, namely rails and the seabirds. There has been only one extinction among the seabirds (arctic tern *Sterna padadisea* Pontoppidan), whereas six new species have colonised Clare Island since 1911 (Chapters 2 and 3).

Finally, as explained, uncertainties remain over the status of the water rail *Rallus aquaticus* Linnaeus, manx shearwater *Puffinus puffinus* (Brünnich) and storm petrel *Hydrobates pelagicus* (Linnaeus) (see Chapters 2 and 3). It is hoped that future surveys will resolve these issues and, though Clare Island is a difficult area in which to carry out a detailed study that investigates all the different breeding bird groups, this manuscript should serve as a broad baseline against which additional data can be compared.

Acknowledgements
This paper could not have been prepared without the generous, and expert, assistance of Brian Caffery of BirdWatch Ireland who provided guidance and access to all the atlas files. We are also grateful to Ricky Whelan for his constructive help and advice. We would also like to express our deepest gratitude to all the fieldworkers who have carried out surveys, including Ian Forsyth and Pamela Allen in 1996 and 1997, and for their ingenious method of summarising the results, and to Richard H. Coombes (2000). We would like to gratefully acknowledge the constructive comments of an anonymous referee.

REFERENCES

Balmer, D., Gillings, S., Caffrey, B., Swann, B., Downie, I. and Fuller, R. 2013 *Bird Atlas 2007–11: the breeding and wintering birds of Britain and Ireland*. Thetford. BTO Books.

Barlee, J. and Ruttledge, R.F. 1945 Notes on the present status of birds on Clare Island. *Irish Naturalists' Journal* **8**, 311–13.

Bauer, P., Thorpe, A. and Brunet, G. 2015 The quiet revolution of numerical weather prediction. *Nature* **525**, 47–55.

Bibby, C.J., Burgess, N.D., Hill, D.A. and Mustoe, S. 2000 *Bird census techniques*. Second Edition. London. Academic Press.

Coombes, R.H. 2000 Clare Island. Birds noted on the island during 3–day visit 29/4 to 1/5/2000. Unpublished Report.

D'Arcy, G. 1992 Fauna observed on Clare Island, Co. Mayo June 14–18 1990. *Irish Naturalists' Journal* **24**, 33–4.

Diamond, J. and May, R.M. 1977 Species turnover rates on islands: dependence on census interval. *Science* **197**, 266–70.

Doyle, G.J. and Foss, P.J. 1986 A resurvey of the Clare Island flora. *Irish Naturalist's Journal* **22**, 85–9.

Forsyth, I. and Allen, P. 1997 *Report to the Royal Irish Academy on the Breeding Birds of Clare Island, Co. Mayo, 1996 and 1997*. Unpublished report to Royal Irish Academy. 9 pages.

Gibbons, D.W., Reid, J.B. and Chapman, R.A. 1993 *The New Atlas of Breeding Birds in Britain and Ireland: 1988–1991*. London. Poyser.

Heaney, L.R. 2000 Dynamic disequilibrium: a long-term large scale perspective on the equilibrium model of island biogeography. *Global Ecology and Biogeography* **9**, 59–74.

Hounsome, M.V. and Rear, D. 1966 The bird population of the Great Blasket Island, Illaunboy and Beginish, Co. Kerry. *Irish Naturalists' Journal* **15**, 169–75.

Hutchinson, C.D. 1989 *Birds in Ireland*. Calton. Poyser.

Kelleher, K.M. and O'Halloran, J. 2006 Breeding biology of the song thrush *Turdus philomelos* in an island population. *Bird Study* **53**, 142–55.

Kelly, T.C. 2000 and 2013 Unpublished survey data.

Lloyd, C.S. 1984 The birds of Clare Island, Co. Mayo in June 1982. *Irish Naturalists' Journal.* **21**, 212–16.

Losos, J.B. and Ricklefs, R. (eds) 2010 *The theory of island biogeography revisited*. Princeton and Oxford. Princeton University Press.

MacArthur, R.H. and Wilson, E.O. 1963 An equilibrium theory of insular zoogeography. *Evolution* **17**, 373–87.

MacArthur, R.H. and Wilson, E.O. 1967 *The theory of island biogeography*. Monographs in Population Biology Volume 1. Princeton. Princeton University Press.

Manne, L.L., Pimm, S.L., Diamond, J.M. and Reed, T.M. 1998 The form of curves: a direct evaluation of MacArthur and Wilson's classic theory. *Journal of Animal Ecology* **67**, 784–94.

McCann, K.S. 2000 The diversity-stability debate. *Nature* **405**, 228–33.

McCollin, D. 2015 The curious case of Skokholm: equilibrium, non-equilibrium and phase shift in an island bird assemblage. *Ecography* **38**, 986–91.

McCollin, D. 2017 Turnover dynamics of breeding landbirds on islands: is island biogeographic theory true but trivial over decadal time scales. *Diversity* **9**, 1–7.

McGeehan, A. 2018 *To the ends of the earth; Ireland's place in bird migration*. Cork. Collins Press.

Pennekamp, F., Pontarp, M., Tabi, A., Altermatt, F., Alther, R., Choffat, Y., Fronhofer, E.A., Ganesanandamoorthy, P., Garnier, A., Griffiths, J.I., Greene, S., Horgan, K., Massie, T.M., Mächler, E., Palamara, G.M., Seymour, M. and Petchey, O.L. 2018 Biodiversity increases and decreases ecosystem stability. *Nature* **563**, 109–12.

Pimm, S.L., Diamond, J., Reed, T.M., Russell, G.J. and Verner, J. 1993 Times to extinction for small populations of large birds. *Proceedings of the National Academy of Sciences, USA* **90**, 10871–5.

Pimm, S.L. and Raven, P. 2000 Extinction by numbers. *Nature* **403**, 843–45.

Reed, T. 1980 Turnover frequency in island birds. *Journal of Biogeography* **7**, 329–35.

Reed, T. 1981 The number of breeding landbird species on British Islands. *Journal of Animal Ecology* **50**, 613–24.

Ricklefs, R.E. and Bermingham, E. 2001 Non equilibrium dynamics of the Lesser Antillian avifauna. *Science* **294**, 1522–24.

Russell, G.J., Diamond, J.M., Pimm, S.L. and Reed, T.M. 1995 A century of turnover: community dynamics on three timescales. *Journal of Animal Ecology* **64**, 628–41.

Russell, G.J., Diamond, J.M., Reed, T.M. and Pimm, S.L. 2006 Breeding birds on small islands: island biography or optimal foraging? *Journal of Animal Ecology* **75**, 324–39.

Ruttledge, R.F. 1950 A list of the birds of the counties of Galway and Mayo: showing their status and distribution. *Proceedings of the Royal Irish Academy* **52B**, 315–81.

Ruttledge, R.F. 1994 *Birds in Counties Galway and Mayo*. Dublin. Irish Wildbird Conservancy.

Sharrock, J.T.R. 1976 *The Atlas of Breeding Birds in Britain and Ireland*. Tring. British Trust for Ornithology.

Simberloff, D. 1974 Equilibrium theory of island biogeography and ecology. *Annual Review of Ecology and Systematics* **5**, 161–82.

Stracey, C.M. and Pimm, S.L. 2009 Testing island biogeography theory with visitation rates of birds to British islands. *Journal of Biogeography* **36**, 1532–9.

Thomson, D.L., Baillie, S.R. and Peach, W.J. 1997 The demography and age specific annual survival of song thrushes during periods of population stability and decline. *Journal of Animal Ecology* **66**, 414–24.

Tilman, D. 1996 Biodiversity: population versus ecosystem stability. *Ecology* **77**, 350–63.

Triantis, K.A., Guilhaumon, F. and Whitaker, R.J. 2012 The island species area relationship: biology and statistics. *Journal of Biogeography* **39**, 215–31.

Ussher, R.J. 1912 A biological survey of Clare Island in the county of Mayo, Ireland and adjoining district. *Proceedings of the Royal Irish Academy* **31**, Sec.11, Part 20.

Williamson, M. 1981 *Island Populations*. Oxford. Oxford University Press.

Williamson, M. 1983 The land-bird community of Skokholm: ordination and turnover. *Oikos* **41**, 378–84.

Williamson, M. 1989 The MacArthur – Wilson theory today: true but trivial. *Journal of Biogeography* **16**, 3–4.

Winters, P. 2006 *A Summer Bird Survey of Clare Island*. Unpublished MSc thesis. Dublin. Centre for the Environment, Trinity College.

CHAPTER 8

SIZE MATTERS: ISLAND AREA AND BREEDING ROOKS *CORVUS FRUGILEGUS* – AN EXAMPLE OF THE ALLEE EFFECT?

T.C. Kelly and P. Smiddy

ABSTRACT

Although nationally abundant and widespread, the rook *Corvus frugilegus* Linnaeus does not breed on Clare Island or on any other of Ireland's offshore islands. Previously, it had been suggested that treeless habitats, such as occur on exposed Atlantic islands, might account for the absence of the rook from these locations. In this paper we set out to explain why the rook does not breed on Ireland's offshore islands. An analysis of the distribution of the rook on the offshore islands of England and Scotland suggests that a threshold area of ≥38,000ha is necessary for the species to establish a viable breeding population. The minimum area required to sustain a rookery exceeds by an order of magnitude the size of the largest offshore Irish island (Inishmore; 3,108ha). It appears that the Allee effect applies to the highly colonial breeding and roosting behaviour of the rook, and that if there is either an insufficiency or a reduction in its preferred grassland feeding habitat, its population experiences inverse density dependence and local extinction.

Introduction

Adult rook. Photo: R.T. Mills

The rook *Corvus frugilegus* Linnaeus is one of the most numerous, familiar and widely distributed bird species in Ireland. For example, Crowe *et al.* (2014) estimated the 'all-Ireland' population at 3.75 million individuals (upper bound = 5.16 and lower bound = 2.5 million). Balmer *et al.* (2013) showed the rook was present in 90% of sampled squares in Ireland during the 2007–11 atlas survey, with breeding confirmed in 79%, compared to 71% in Great Britain. Moreover, the densities, and consequently the biomass of breeding rooks are greater in Ireland than in the strongholds of Scotland and England (Castle 1977; Hutchinson 1989; Chapter 6). Nevertheless, the rook does not nest on Clare Island and it is surprising that despite its numerical abundance and geographical range in this country, it does not breed on any Irish offshore island. However, there is one nineteenth-century record of breeding by rooks on Lambay Island, Co. Dublin (Hart 1883). It was stated that 'there is little timber on Lambay; a small planting of sycamores at the castle', and that

Lassau Wood at Portlea. Note the dense growth of heather and bracken in the foreground. The likely history of this wood that was present during the 1909–11 survey, and its connection with the 'anomalous' Lough Avullin, is comprehensively described in a fascinating review by Feehan (2019). The wood may have formed in the 1850s. Photo: T. Kelly

Lassau Wood interior. The vegetation in the wood is too dense to allow a transect to be walked through its centre and so a point count was conducted on 19 June 2013. The main tree species are 'birch and grey willow never more than 5m high' (Feehan 2019). Willow warbler (N=4 including one carrying food) and one each of robin, wren, blackbird and chaffinch was all that was noted in the 10-minute point count. Photo: T. Kelly

The canopy of Lassau Wood at Portlea. We believe contra Feehan (2019) that this wood could hold breeding rooks *Corvus frugilegus* Linnaeus if all other conditions necessary for the formation of a colony were met. The sparrowhawk *Accipiter nisus* (Linnaeus) bred successfully in this wood in the 1990's (Bob Cussen personal observations). Photo: T. Kelly

there was 'a good deal of cultivation at the south and south-west of the island' (Hart 1883). There are no recent records of breeding there in the literature (Hutchinson 1975).

The rook, however, does breed on the Isle of Man, and on some of the larger islands off the coasts of England and Scotland. The existence of rook colonies on these offshore archipelagos and their absence from others, suggests that there is a relationship between the probability of the rook being present during the breeding season and the area of the island.

However, as we have already seen, the relationship between island area and species richness is generally linear, although the reasons for this have been long and intensely debated (e.g. Abbott 1980; Losos and Ricklefs 2010; and references therein). Interestingly, there appears to have been few studies that have addressed the issue of area from the standpoint of one species. As already shown, the Corvidae as a family are well represented in the avifaunas of Irish islands (see Chapter 6; Pimm *et al.* 1993), with the jay *Garrulus glandarius hibernicus* Witherby and Hartert and the rook being the only species which are absent as breeding birds. The jay remains relatively scarce in Ireland (e.g. Balmer *et al.* 2013) and does not breed on the Isle of Man (Sharpe 2007; Balmer *et al.* 2013) due in part, presumably, to its specialised deciduous woodland habitat requirements. However, it has also been shown that small populations of crows can be highly persistent (Pimm *et al.* 1993) though the rook was not included in that analysis.

The rook is also an unusual species in that it habitually breeds in colonies and its marked colonial behaviour sets it apart from all other corvid species breeding in Ireland (e.g. Coombs 1978; Cramp and Perrins 1994). The colonial habit, however, implies a behavioural requirement as well as an ecological one, and, in combination, it may prove difficult to satisfy both the nutritional, and social needs of this species. Perrins (1985) points out, the rook 'feeds almost exclusively on areas of short grass' in 'large flocks which must have a considerable food requirement'. Interestingly, Perrins (1985) also comments that it seems possible that the rook did not exist in north-western Europe before man opened the dense woodland habitats, concluding that 'perhaps it was a bird of the wooded steppes (it uses trees for roosting and nesting)'.

While the main aim of this manuscript is to find an explanation for the absence of the rook as a breeding bird from Clare Island, and indeed from all other islands off the coast of Ireland, it is also our objective to identify some of the properties of those offshore insular locations where the rook has established colonies.

Study area and methods

The study area is the suite of offshore islands of Ireland and Great Britain. As with previous studies, we omitted islands that are either very close to the mainland, e.g. Sherkin Island, Co. Cork, or are connected by a bridge, e.g. Valentia Island, Co. Kerry and Achill Island, Co. Mayo, Anglesey Island in Wales and the Isle of Skye in Scotland. We also omitted inshore islands such as those located in the Firth of Clyde or in Bantry Bay, Co. Cork.

Presence of breeding rooks on offshore islands

We consulted all available literature relating to the islands of Ireland and Great Britain including Ussher and Warren (1900), Kennedy *et al.* (1954), Ruttledge (1966), Thom (1986), Elliott (1989), Hutchinson (1989), Lovegrove *et al.* (1994), Brown and Grice (2005), Forrester *et al.* (2007), Sharpe (2007), Balmer *et al.* (2013), and monographs such as Lysaght (2002) and Berry (2009). In addition, we reviewed data gathered during the most recent rookery surveys of Great Britain and Northern Ireland (Sage and Whittington 1985; Marchant and Gregory 1999).

Areas of islands

Since distance from the mainland is clearly not a variable, and rooks occur on Shetland, the Orkneys and the Outer Hebrides, but are absent from many islands much closer to the coastline of Scotland, Wales, England and Ireland, it appears that area may explain the observed distribution of the rook on offshore islands.

Data on the offshore islands of Ireland were obtained for individual islands from a range of sources (e.g. Ussher 1912; Sharrock 1973; Lysaght 2002; Berry 2009). The areas of the islands off the coast of Scotland were obtained from Hewell-Smith (2004) and off the English and Welsh islands from Berry (2009). All areas are given in hectares. The statistical analysis was undertaken using Graphpad PRISM version 8.3.1.

Results

Rooks on offshore islands in Ireland and Great Britain

Rooks do not breed on any of the offshore islands of Ireland. Likewise, there are no rookeries on any island off the coast of Wales, and only one island, the Isle of Wight, holds breeding rooks off the coast of England (Balmer *et al.* 2013). There are several rookeries on the Isle of Man. Rooks do not breed on the Scilly Islands; although a rookery was present between 1928 and 1975 on Jersey in the Channel Islands (Marchant and Gregory 1999; Young *et al.* 2014).

However, rooks currently breed on at least six of Scotland's unconnected offshore islands (Table 1). Interestingly, one of these is Iona which, because it lies close to the western extremity of the Isle of Mull, should properly be considered a 'satellite' of the colonies (n = 2, Castle 1977) on this very large Inner Hebridean island. In addition, Castle (1977) noted colonies on Hoy and Rousay in the Orkney group of islands. It appears that Hoy and Rousay are not regularly occupied by breeding rooks and the same appears to be true of the Isle of Jura (see Balmer *et al.* 2013).

Areas of offshore islands occupied by rooks

The mean area of all offshore islands currently occupied by breeding rooks (Iona excluded) is 87,684ha ±23,014, SE, but the range is relatively enormous at 179,798ha (Table 2), the skewness being due to the massive size of Lewis and Harris. The median area and the ±95% Confidence Intervals are shown in Fig. 1. The key figure here, however, is the minimum area of an island currently occupied by rooks at 38,100ha, which is the Isle of Wight in southern England. The smallest occupied Scottish island is Mainland in

Table 1
Offshore islands in Britain that currently hold breeding rooks.

Country	Island name	Area in ha
England	Isle of Wight	38,110
Isle of Man	Isle of Man	58,793
Scotland	Mainland – Shetland	96,879
	Mainland – Orkney	52,325
	Mull (including Iona)	88,712
	Islay	61,956
	Lewis and Harris	217,898

the Orkneys (52,325ha, Table 1). It should be noted that rooks have bred on Jersey 11,630ha but not on Arran (42,201ha).

However, if Iona is included, the average area of those islands with breeding rooks is 76,833ha ±22,693, SE; but the minimum area drops to 877ha (Table 2), although the range is even larger at 217,021ha. Finally, if the areas of all offshore islands which have held breeding rooks are pooled, including Hoy, Rousay and Jersey in the Channel Islands, the mean area is 56,775ha ±17,204, SE (Table 2).

Discussion

It has long been known that the rook does not breed on any offshore island in Ireland (Ussher and Warren 1900; Ussher 1912; Kennedy *et al.* 1954; Ruttledge 1966; Hutchinson 1989; Lysaght 2002). Breeding rooks were not recorded in the first Clare Island survey; the rook 'does not breed on Clare Island, but considerable numbers resort thither in the end of June and onwards until March' (Ussher 1912). Later, Ussher remarked that Robert Lloyd Praeger saw '100 together in July' and he notes that 'they feed even on the mountain moors; and young rooks can be seen on the island fed by their parents' (Ussher 1912). Likewise, Sharrock (1973) reported that rooks were regular visitors to Cape Clear Island, Co. Cork, but 'depart eastwards at dusk'.

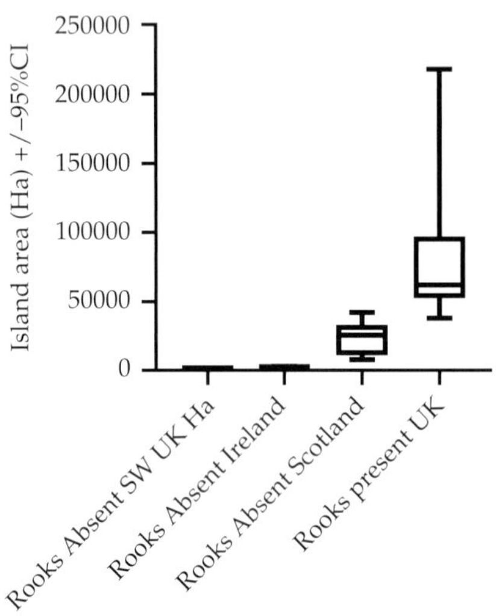

Fig.1 Median area of islands with and without breeding rooks and the Confidence Intervals. The south west UK islands (N = 6) include Skomer, Puffin and Walney islands; the Irish islands N = 6 include those with the largest area incuding Inishmore, Rathlin Island, Clare Island and Arranmore. The islands without rooks off the coast of Scotland include North and South Uist, Yell, Tiree and the Isle of Arran. Those UK islands (N = 7) with breeding rook colonies are listed in Table 1; the median is 61,156ha the upper bound is 217,898ha and the lower bound is 31,800ha. The mean is 87,684ha.

These observations show that the rook can reach the offshore islands but does not breed there and the question addressed in this paper is: why not? Moreover, in the light of the acquisition of much recent data, their absence is even more striking now than it was in the past, as the sequence of breeding bird atlases and other surveys have shown how abundant and widespread this species is in Ireland (Sharrock 1976; Gibbons *et al.* 1993; Balmer *et al.* 2013; Crowe *et al.* 2010, 2014).

Ussher and Warren (1900) stated that rooks do not 'resort in the breeding-season' 'to the islands, unless these contain trees' and Kennedy *et al.* (1954) noted that 'from the treeless marine islands the rook is absent as a breeding species', while Ussher (1912) remarked that the rook 'nests as far as trees extend on the mainland'. The implication of these observations is that the absence of trees is the only resource that prevents the rook from nesting on offshore islands in Ireland. Of course, trees are the preferred breeding site for this species in which to form its colonies and a supply of small branches and twigs is essential to enable them to build their substantial nests. But while some large islands such as the Great Blasket (Kerry) remain treeless, there are plantations on, for example, Cape Clear, Lambay (Dublin) and Clare Island, of a height and size that would be suitable for the formation of a rookery. Furthermore,

Table 2
Mean, minimum, maximum and range of island areas with breeding rooks.

Islands with rooks both present and past	Mean area ha	Minimum area ha	Maximum area ha	Range ha
All islands with rooks – excluding Iona	87,684	38,100	217,898	179,798
All islands with rooks – including Iona	76,833	877	217,898	217,021
All islands with rooks present and past	56,775	877	217,898	217,021

other corvid species including magpie *Pica pica* (Linnaeus), hooded crow *Corvus cornix* Linnaeus and raven *Corvus corax* Linnaeus as well as chough *Pyrrhocorax pyrrhocorax* (Linnaeus) appear to have little difficulty in finding material to build their nests. Thus, while in the past the absence of trees may have been a critical factor in preventing rooks from breeding on offshore islands such as Cape Clear and Clare Island, it would not appear to be the case today. We do not agree that Lassau Wood is insufficient to hold nesting rooks (Feehan 2019).

The rook is an obligate colonial species—a behavioural trait that is unique among the Corvidae breeding in Ireland. However, much remains to be learned about the precise set of circumstances that favour the establishment of a colony among bird species generally (Brown and Rannala 1995; Danchin and Wagner 1997; Danchin *et al.* 2004; Valone 2007) and particularly in the rook (Griffin 1999; Griffin and Thomas 2000). Brown and Rannala (1995) contend that obligate colonial bird species do not appear to obey the predictions of the Ideal Free Distribution model developed by Fretwell and Lucas (1970), but instead conform to the terms of the Allee effect, i.e. a positive relationship between fitness (i.e. survival and reproductive success) and the numbers and densities of conspecifics (Stephens and Sutherland 1999). In contrast, colonial species, at low numbers, may experience inverse density dependence possibly leading to local extinction (Courchamp *et al.* 1999). The Allee effect of reduced survival and reproductive success occurs, counter-intuitively, when populations of certain species decline. The relaxation of density dependence would be expected to result in enhanced fitness with higher compensatory reproductive success due to their being less competition for resources (Stephens and Sutherland 1999; Berec *et al.* 2007), but this does not happen in some social species, e.g. the herring gull *Larus argentatus* Pontoppidan on Cape Clear Island (Crème 1997). These effects are relevant to the presence or absence of rook colonies on offshore islands.

Island area is likely to be the main constraint on colonisation and long-term occupancy by breeding rooks. To breed successfully, rooks will need a sufficiently large foraging area, adjacent to, or within 1 to 2km of the colony (Macdonald and Whelan 1986; Mason and Macdonald 2004) in which to provision themselves and their chicks. Clutch size in the rook averages around four eggs, and when the young hatch the brood will occupy the nest for 30 to 36 days, after which the fledged juveniles may be fed by the adults for a further six weeks (Cramp and Perrins 1994). It seems to be generally agreed that the rook requires a large area of pasture in which to forage during the breeding season (Griffin and Thomas 2000; Mason and Macdonald 2004) though the food subsidy provided by refuse tips may also be exploited (Olea and Baglione 2008; Olea 2009). However, it is interesting to note that 'In contrast to many colonial birds rooks have a strong association with their nesting rookeries throughout the year' (Griffin and Thomas 2000) and breeding in winter has been recorded (Coombs 1978). This suggests that an estimate of the carrying capacity for a rook colony would have to include a sufficiently large area of foraging habitat to enable its members to provision themselves throughout the year (Olea and Baglione 2008).

A consequence of the Allee effect is that survival and reproductive success will increase with an expansion in the numbers of birds in the colony; but this in turn will require increasing areas of suitable habitat. And, conversely, if there is a decrease in available habitat, the numbers of birds in the colony will fall leading possibly to inverse density dependence and possible extinction. For example, Orłowski and CZapulak (2007) show in their detailed study that the probability of abandonment of a rookery is 'strongly' negatively correlated with its size and that 'the median for abandoned colonies was 16 nests'.

Our estimates suggest that rook colonies only occur on offshore islands with a minimum area ≥38,000ha. While there is a rook colony on Iona (877ha) it is assumed here that it survives because of the adjacent large island of Mull on which there are two colonies (Castle 1977). Rooks also breed on Lismore Island (2,351ha), Great Cumbrae (1,168ha) and Bute (12,217ha) in the Firth of Clyde, but these inshore islands appear to be flanked by the mainland where presumably there is readily available foraging habitats. Moreover, the more isolated and exposed Isle of Arran (43,201ha) is without breeding rooks, as are both North (30,305ha) and South Uist (32,026ha); similarly, rooks are absent from Tiree (7,834ha), Unst (12,068ha) and Yell (21,211ha).

This suggests that there is a threshold amount of habitat (e.g. Lande 1987; Fahrig 2001) which must exist before the rook can successfully colonise and sustain its autonomous occupancy of an island. The threshold is likely to include a sufficiently large area of grassland foraging habitat and woodland in order to allow the establishment of a rookery large enough—probably greater than 16 nests—to be buffered against inverse density dependent 'Allee effect' extinctions. The rook may also experience competition on marine islands from species like the chough and gulls (particularly the common gull *Larus canus* Linnaeus and lesser black-backed gull *Larus fuscus* Linnaeus), which are now known to feed extensively on terrestrial invertebrates (e.g. Crème 1997; Kelly *et al*. 2012).

We suggest that the constraints imposed by the Allee effect on this highly social corvid species preclude its ability to permanently colonise an island below a threshold minimum area (see also Taylor and Hastings 2005).

The data presented here suggests that the threshold island area may be of the order of 38,000ha, but with the caveat that 'area' alone may not meet the rook's nutritional and behavioural demands. It is the combined area of grassland and woodland habitat that specifically meets the rook's ecological requirements that matters. Climate change may also be an influence as some of the offshore islands have been colonised relatively recently associated with an expansion in the geographical range of the rook (Venables and Venables 1953; Venables 1973; Williamson 1975).

Summary

The rook may have arrived more recently in Britain and Ireland than the other corvid species, as woodland clearances creating large areas of grassland (the rook's preferred habitat) were necessary before it could become established as a resident species. It is a colonial species which, unusually, appears to be dependant for much of the year on habitats in which it can feed within 1 to 2km of the rookery. It is probable that because of its colonial breeding and gregarious feeding and roosting behaviour, it exhibits positive density dependence in survival and reproductive success typical of the Allee effect. Equally however, a decline in the size of a rookery could lead to inverse density dependence and local extinction. Consequently, there is likely to be a threshold (i.e. a minimum area of feeding and nesting habitat) in area of an offshore island below which the rook could not establish a viable breeding colony. We suggest the threshold may be as large as 38,000ha for offshore marine islands situated more than 2km from the nearest land. No island—including Clare Island—off the coast of Ireland exceeds the minimum threshold, and consequently despite its numerical abundance and widespread distribution in this country the rook does not breed on any marine offshore island.

Acknowledgements

We are indebted to Dr Tom Reed for his constructive comments on an earlier draft of this paper and to those of an anonymous referee.

REFERENCES

Abbott, I. 1980 Theories dealing with the ecology of landbirds on islands. *Advances in Ecological Research* **2**, 329–71.

Balmer, D.E., Gillings, S., Caffrey, B.J., Swann, R.L., Downie, I.S. and Fuller, R.J. 2013 *Bird atlas 2007–11: the breeding and wintering birds of Britain and Ireland*. Thetford. BTO Books.

Berec, L., Angulo, E. and Courchamp, F. 2007 Multiple Allee effects and population management. *Trends in Ecology and Evolution* **22**, 185–91.

Berry, R.J. 2009 *The natural history of islands*. London. Collins.

Brown, A. and Grice, P. 2005 *Birds in England*. London. Poyser.

Brown, C.R. and Rannala, B. 1995 Colony choice in birds: models based on temporally invariant site quality. *Behavioral Ecology and Sociobiology* **36**, 221–8.

Castle, M.E. 1977 Rookeries in Scotland – 1975. *Scottish Birds* **9**, 327–34.

Coombs, F. 1978 *The crows*. London. Batsford.

Courchamp, F., Clutton-Brock, T. and Grenfell, B. 1999 Inverse density dependence and the Allee effect. *Trends in Ecology and Evolution* **14**, 405–10.

Cramp, S. and Perrins, C.M. 1994 *The birds of the Western Palearctic*. Volume 6. Oxford. Oxford University Press.

Crème, G. 1997 Aspects of the biology of an expanding population of lesser black-backed gulls *Larus fuscus* in southwest Ireland. Unpublished Ph.D. thesis. Cork. University College.

Crowe, O., Coombes, R.H., Lysaght, L., O'Brien, C., Choudhury, K.R., Walsh, A.J., Wilson, J.H. and O'Halloran, J. 2010 Population trends of widespread breeding birds in the Republic of Ireland 1998–2008. *Bird Study* **57**, 267–80.

Crowe, O., Musgrove, A. and O'Halloran, J. 2014 Generating population estimates for common and widespread breeding birds in Ireland. *Bird Study* **61**, 82–90.

Danchin, E., Giraldeau, L-A., Valone, T.J. and Wagner, R.H. 2004 Public information: from noisy neighbours to cultural evolution. *Science* **305**, 487–91.

Danchin, E. and Wagner, R.H. 1997 The evolution of coloniality: the emergence of new perspectives. *Trends in Ecology and Evolution* **12**, 342–7.

Elliott, R.E. 1989 *The birds of Islay*. London. Christopher Helm.

Fahrig, L. 2001 How much habitat is enough? *Biological Conservation* **100**, 65–74.

Feehan, J. 2019 *Clare Island*. Dublin. Royal Irish Academy.

Forrester, R., Andrews, I., McInerney, C., Murray, R., McGowan, R., Zonfrillo, B., Betts, M., Jardine, D. and Grundy, D. (eds) 2007 *The birds of Scotland*. Aberlady. Scottish Ornithologists' Club.

Fretwell, S.D. and Lucas, H.L. Jr. 1970 On territorial behaviour and the factors influencing habitat distribution in birds. 1. Theoretical development. *Acta Biotheoretica* **19**, 1–36.

Gibbons, D.W., Reid, J.B. and Chapman, R.A. (eds) 1993 *The new atlas of breeding birds in Britain and Ireland: 1988–1991*. London. Poyser.

Griffin, L.R. 1999 Colonization patterns of rook *Corvus frugilegus* colonies: implications for survey strategies. *Bird Study* **46**, 170–3.

Griffin, L.R. and Thomas, C.J. 2000 The spatial distribution and size of rook (*Corvus frugilegus*) breeding colonies is affected by both the distribution of foraging habitat and by intercolony competition. *Proceedings of the Royal Society* **267B**, 1463–7.

Hart, H.C. 1883 The birds of Lambay Island, Co. Dublin. *Zoologist* (3) **7**, 155–64, 225–6.

Hewell-Smith, H. 2004 *The Scottish Islands*. Edinburgh. Canongate.

Hutchinson, C.D. (ed.) 1975 *The birds of Dublin and Wicklow*. Dublin. Irish Wildbird Conservancy.

Hutchinson, C.D. 1989 *Birds in Ireland*. Calton. Poyser.

Kelly, M., Oxbrough, A. and Kelly, T.C. 2012 Some observations on the diet of adult common gulls (*Larus canus*) at a coastal breeding colony in County Kerry. *Irish Birds* **9**: 357–60.

Kennedy, P.G., Ruttledge, R.F. and Scroope, C.F. 1954 *The birds of Ireland*. Edinburgh and London. Oliver and Boyd.

Lande, R. 1987 Extinction thresholds in demographic models of terrestrial populations. *American Naturalist* **130**, 624–35.

Losos, J.B. and Ricklefs, R. (eds) 2010 *The theory of island biogeography revisited*. Princeton and Oxford. Princeton University Press.

Lovegrove, R., Williams, G. and Williams, I. 1994 *Birds in Wales*. London. Poyser.

Lysaght, L. 2002 *An atlas of the breeding birds of the Burren and Aran Islands*. Dublin. BirdWatch Ireland.

Macdonald, R. and Whelan, J. 1986 Seasonal variation in feeding range and flock structure of the rook (*Corvus frugilegus*) in eastern Ireland. *Ibis* **128**, 540–47.

Marchant, J.H. and Gregory, R.D. 1999 Numbers of nesting rooks *Corvus frugilegus* in the United Kingdom in 1996. *Bird Study* **46**, 258–73.

Mason, C.F. and Macdonald, S.M. 2004 Distribution of foraging rooks *Corvus frugilegus* and rookeries in a landscape in eastern England dominated by winter cereals. *Folia Zoologica* **53**, 179–88.

Olea, P.P. 2009 Analysing spatial and temporal variation in colony size: an approach using autoregessive mixed models and information theory. *Population Ecology* **51**, 161–74.

Olea, P.P. and Baglione, V. 2008 Population trends of rooks *Corvus frugilegus* in Spain and the importance of refuse tips. *Ibis* **150**, 98–109.

Orłowski, G. and CZapulak, A. 2007 Different extinction risks of the breeding colonies of rooks *Corvus frugilegus* in rural and urban areas of SW Poland. *Acta Ornithologica* **42**, 145–55.

Perrins, C.M. 1985 The effect of man on the British avifauna. In Bunning, J.L. (ed) *Proceedings of the Birds and Man Symposium*. 5–29. Johannesburg.

Pimm, S.L., Diamond, J., Reed, T.M., Russell, G.J. and Verner, J. 1993 Times to extinction for small populations of large birds. *Proceedings of the National Academy of Sciences, USA* **90**, 10871–5.

Ruttledge, R.F. 1966 *Ireland's birds*. London. Witherby.

Sage, B. and Whittington, P.A. 1985 The 1980 sample survey of rookeries. *Bird Study* **32**, 77–81.

Sharpe, C.M. 2007 *Manx bird atlas*. Liverpool. Liverpool University Press.

Sharrock, J.T.R. 1973 *The natural history of Cape Clear Island*. Berkhamsted. Poyser.

Sharrock, J.T.R. 1976 *The atlas of breeding birds in Britain and Ireland*. Tring. British Trust for Ornithology.

Stephens, P. and Sutherland, W. 1999 Consequences of the Allee effect for behaviour, ecology and conservation. *Trends in Ecology and Evolution* **14**, 401–5.

Taylor, C.M. and Hastings, A. 2005 Allee effects and biological invasions. *Ecology Letters* **8**, 895–908.

Thom, V.M. 1986 *Birds in Scotland*. Calton. Poyser.

Ussher, R.J. 1912 A biological survey of Clare Island in the county of Mayo, Ireland, and the adjoining district. *Proceedings of the Royal Irish Academy* **31**, Sec.11, Part 20.

Ussher, R.J. and Warren, R. 1900 *Birds of Ireland*. London. Gurney and Jackson.

Valone, T.J. 2007 From eavesdropping on performance to copying the behaviour of others: a review of public information use. *Behavioral Ecology and Sociobiology* **62**, 1–14.

Venables, L.S.V. 1973 Rooks in Shetland. *Scottish Birds* **7**, 406.

Venables, L.S.V. and Venables, U.V. 1953 Rooks nesting in Shetland. *British Birds* **46**, 265–66.

Williamson, K. 1975 Birds and climate change. *Bird Study* **22**, 143–64.

Young, G., Lawlor, M., Paintin, T., Atkinson, M. and Watson, J. 2014 *A working list of the birds of the Channel Islands*. www.jerseybirds.co.uk. Accessed 19 September 2016.

SUMMARY

T.C. Kelly

Roonagh Pier from where most visitors to Clare Island depart from the mainland. The adjacent coastline has several marshes where dunlin *Calidris alpina* (Linnaeus) breed – although the species has never been recorded on Clare Island. Many rare species including a royal tern/west African crested tern *Thalasseus maxima/albididorsalis* (Boddaert) have been recorded in areas nearby this pier (Hobbs, J. *A list of Irish birds*. Version.8, 2020) and Joe Doolan's www.irishbirding.com. Photo: T. Kelly

Clare Island was formed about 8,000 years ago (yBP). The avifauna was first surveyed in the 1909 to 1911 interval and not again until 1945. Thereafter, the island's birds were investigated by the various breeding and winter atlas surveys together with assorted expeditions of short duration (details in Chapters 2 and 7).

The systematic list (Chapter 2) includes 127 species but there is uncertainty about some, e.g. the Eurasian wigeon *Anas penelope* Linnaeus, which has not been seen since the first survey. Some species have become extinct, e.g. corn bunting *Emberiza calandra* (Linnaeus), common whitethroat *Sylvia communis* Latham and the golden eagle *Aquila chrysaetos* (Linnaeus). New arrivals include the tree sparrow *Passer montanus* (Linnaeus), chaffinch *Fringilla coelebs* Linnaeus and the willow warbler *Phylloscopus trochilus* (Linnaeus). The inventory of species is relatively short especially when compared with those of the bird observatory on Cape Clear and the well watched Great Saltee Island. The absence of a muddy shoreline explains why common wading bird species such as the dunlin *Calidris alpina* (Linnaeus), greenshank *Tringa nebularia* (Gunnerus) and grey plover *Pluvialis squatarola* (Linnaeus) have never been recorded on Clare Island. Likewise, the great tit *Parus major* Linnaeus, blue tit *Cyanistes caeruleus* (Linnaeus), bullfinch *Pyrrhula pyrrhula* (Linnaeus) and spotted flycatcher *Muscicapa striata* (Pallas) remain to be detected. The ring ouzel *Turdus torquatas* Linnaeus is formally removed from the inventory as there is no reliable evidence of it having been seen on Clare Island—even though it is frequently recorded, although a scarce passage migrant, elsewhere on the Irish coastline.

The seabirds of Clare Island (Chapter 3) have been thoroughly surveyed, particularly over the past 20 years. Currently, 14 species of seabird breed on Clare Island compared to the eight noted in 1912 by Ussher and his co-workers. One species, the arctic tern *Sterna paradisaea* Pontoppidan, is no longer present, whereas the fulmar *Fulmarus glacialis* (Linnaeus), common gull *Larus canus* Linnaeus, lesser black-backed gull *Larus fuscus* Linnaeus, great cormorant *Phalacrocorax carbo* (Linnaeus), gannet *Morus bassanus* (Linnaeus) and, most recently (2015), the great skua *Stercorarius skua* (Brünnich)[1], have established breeding populations on the island at various times over the past century. In fact, remarkably, the species richness of seabirds on Clare Island has increased by 67 to 75% since 1912. Despite a marked recent decline of over 1,500 apparently occupied sites (AOS) the fulmar is the most numerous of the island's seabirds and its colony is the third largest of its kind in Ireland. Other species have also decreased, including the shag *Phalacrocorax aristotelis* (Linnaeus), kittiwake *Rissa tridactyla* (Linnaeus) and common gull, though probably in each case for different reasons. The gannet population has undergone a spectacular recent increase and now numbers some 352 breeding pairs. The 1999 survey showed the presence of 15,000+ individual seabirds of which over half were fulmars, whereas the 2015–17 census yielded a total of between 10,000 and 11,000 seabirds—a loss over the 16–17-year interval of in excess of 4,000 individual birds or approximately 30%. Moreover, the fulmar population was now less than half, indicating the severity of the decline in the numbers of that species. The estimated biomass of the 1999 seabird population was approximately 12,040kg. On the basis that the 1999 population would consume some 30% of their body weight per day, the biomass of prey consumed per day would be of the order of 3,612kg. Although the population has decreased by 30% since 1999, the overall biomass has declined by 807kg or 7% as some of the largest species such as the gannet and great black-backed gull *Larus marinus* Linnaeus have increased over the interval. Based on the current estimate of approximately 11,000 seabirds on Clare Island, it appears that these may consume 3,321kg of prey per day.

Some uncertainty remains over the breeding status of the storm petrel *Hydrobates pelagicus* (Linnaeus) and manx shearwater *Puffinus puffinus* (Brünnich) and while both are suspected to be present on Clare Island no confirmatory evidence has been obtained.

The pattern of migratory movements of seabirds around Clare Island remains unstudied and, for example, the Leach's petrel *Oceanodroma leucorhoa* (Vieillot), Sabine's gull *Xema sabini* (Sabine) and sooty shearwater *Puffinis griseus* (Gmelin) have not been recorded to date. This is unquestionably due to a lack of effort, compared to say the Bridges of Ross in Co. Clare and Kilcummin Head in Co. Mayo.

The breeding land birds of Clare Island (Chapter 4) were comprehensively surveyed by the transect method; data were collected on 23 walked transects carried out in mid-June 2013 and analysed using the program Distance. The aim was to provide a preliminary estimate of the numerical abundance of the breeding land bird species on Clare Island. A total of 28 species (inclusive of the pheasant *Phasanius colchicus* Linnaeus and common kestrel *Falco tinnunculus* Linnaeus) was recorded, accounting for approximately 20% of the total number of species breeding on the island of Ireland. Only four migrant land birds breed, approximately 14% of the total number of the land bird taxa on the island. The island land bird community was estimated to comprise ≥8,000 individuals and was dominated

[1] The great skua did not breed on Clare Island in 2019 (Eoin McGreal personal observation).

by open-habitat specialist and generalist species, e.g. meadow pipit *Anthus pratensis* (Linnaeus), northern wheatear *Oenanthe oenanthe* (Linnaeus), wren *Troglodytes troglodytes* (Linnaeus), dunnock *Prunella modularis* (Linnaeus), robin *Erithacus rubecula* (Linnaeus) and blackbird *Turdus merula* Linnaeus. Despite the presence of some woodland habitat, many hole-nesting species, e.g. the family Paridae, were absent, as was the goldcrest *Regulus regulus*[2] (Linnaeus) and treecreeper *Certhia familiaris* Linnaeus. Only two *Carduelis* and one *Emberiza* species were found to be breeding. There are four crow species breeding on Clare Island, but the rook *Corvus frugilegus* Linnaeus is absent. The land bird community on Clare Island appears to be unique in Ireland due to the exceptionally high numbers of wheatear breeding there. The breeding land bird community is dominated by species with a clutch size of four eggs, and those that produce two broods per annum. In terms of trophic relationships, predators of invertebrates accounted for over 95% of the island's total number of land birds. Herbivorous species, by contrast, were remarkably scarce and only amounted to ≤4% of the total. The estimated total biomass of land birds was ≥376.7kg or ≥24kg per km^2. The hooded crow *Corvus cornix* Linnaeus is the 'heaviest' terrestrial bird species on the island.

The waterbirds of Clare Island (Chapter 5) include five wading bird species that breed on the island, the same number as was discovered during the first survey conducted in 1909–11. Thus, the oystercatcher *Haematopus ostralegus* Linnaeus, ringed plover *Charadrius hiaticula* Linnaeus, northern lapwing *Vanellus vanellus* (Linnaeus), common snipe *Gallinago gallinago* (Linnaeus) and common sandpiper *Actitis hypoleucos* (Linnaeus) are all present on the island today as they were over 100 years ago. This number of wading bird taxa represents 50% of the current species richness of Charadriidae and Scolapacidae which regularly breed in Ireland. Currently, none of the species is numerically abundant and this also seems to have been the situation during the first survey, despite the absence of most of the mammalian predators known to occur in Ireland; typical avian predators are of course present as they are elsewhere in Ireland. However, waterbodies are surprisingly few on Clare Island and the largest—Lough Avullin—has changed from an open water lake habitat of some 1.5ha in 1909, to an impenetrable reed swamp, over the 100+ year interval since the first survey. This may explain why Clare Island holds only one duck species—the mallard *Anas platyrhynchos* Linnaeus—and only one rail—the moorhen *Gallinula chloropus* (Linnaeus)—though the latter only breeds there sporadically. The water rail *Rallus aquaticus* Linnaeus, which appears to have been relatively common during the time of the first expeditions, has not been recorded since, though the large and dense reed bed at Lough Avullin looks an ideal breeding habitat for this species, but one that is exceptionally difficult to survey. Overall, the low diversity of water birds on Clare Island shows that island area *per se* is a poor predictor of species richness in this group. The island looks like an ideal location for the more terrestrial rare Nearctic waders including the buff-breasted sandpiper *Tryngites subruficollis* (Vieillot), American golden plover *Pluvialis dominica* (P.L.S. Müller) and the pectoral sandpiper *Calidris melanotos* (Vieillot), though none has been recorded there to date probably due to the low number of ornithologists visiting the island during the autumn migration time. A dotterel *Charadrius morinellus* Linnaeus was seen by Eoin McGreal on 11 April 2011 near the signal tower.

In the study of breeding land birds on the offshore islands of Ireland (Chapter 6), apparently the first of its kind, we reviewed the published information on the breeding land birds (raptors excluded) of 29 offshore Irish islands. Fifty-seven passerine and near passerine species have been reported as breeding on Irish offshore islands. Thus some 86% of the land bird species breeding on the island of Ireland also breed on the offshore islands. A frequency distribution shows that some species are occupants of all—rock pipit *Anthus petrosus* (Montagu)—or most—wheatear—of the islands, whereas endemic subspecies such as the dipper *Cinclus cinclus* (Linnaeus) and coal tit *Periparus ater* (Linnaeus) have occurred on only one of the islands in the sample. Migrants number 15 species, approximately 27% of the total number of species on the totality of the islands. There are some notable absentees including the treecreeper,[3]

[2] Goldcrest bred in 2006 (Winters 2006) but not subsequently.

[3] Seen recently – Patsy Winters personal observation.

jay *Garrulus glandarius* (Linnaeus) and bullfinch. While the absence of the first two species is probably explained by the absence of woodland habitats on most, but not all, of the islands, the failure of the bullfinch to establish breeding populations is more problematic.

An analysis of the species-area relationship was, unsurprisingly, positive, and statistically highly significant. An NMDS ordination analysis also showed area, latitude, the presence of freshwater lentic habitat and human habitations past or present to be significant influences on species richness. An aim of the study was to expand the concept of 'pool size' as developed by several authors. Thus, the frequency of occurrence of 57 land bird species on the 29 islands (*Y)* was regressed in Log_{10} by Log_{10} plots against the overall abundance of their populations on the island of Ireland as a whole (*N*), as well as their total biomass (*M*) (*M*=*N* x bird body weight) and the extent of their geographical distribution (*D*), i.e. the total number of occupied breeding atlas squares. Moreover, based on an analysis of the relationship between *N* and *D* on the island of Ireland, and between bird species weight and *D*, it was hypothesised that the number of occupied islands would be correlated with *N* and not with bird weight and therefore, *M*. However, despite the obvious influence of area, the results also show that, contrary to expectation, there is no relationship between *N* and the number of occupied islands (*Y*), whereas there is a relationship between *M* and *Y*. These results suggest that there is a higher proportion of heavier bird species on offshore islands than on the island of Ireland as a whole, and that those species which, numerically speaking, are relatively scarce on the 'mainland', e.g. rock pipit occur on a high proportion of the offshore islands. This probably explains why there is no discernible relationship between *N* and *Y*.

The equilibrium number of land bird species on Clare Island (Chapter 7) was also studied. This study had two aims: (i) to estimate the equilibrium number of breeding land bird species on Clare Island and (ii) to describe the apparent turnover of species *T* since the findings of the first survey (1909–11). Although 35 land bird species are known to have bred at least once over the 104-year interval, the equilibrium number of species (i.e. where the curve of extinctions and immigrations intersect) is estimated to be 24. This, interestingly, is four less than the total number of species (n=28) listed by Ussher in the results of the first survey. There have been seven extinctions (*E*) and six immigrations (*I*) since the first survey. Extinct species include the corn bunting, dipper, common whitethroat, sand martin *Riparia riparia* (Linnaeus), common cuckoo *Cuculus canorus* Linnaeus and twite *Carduelis flavirostris* (Linnaeus). Immigrants include the willow warbler, tree sparrow and chaffinch. The rate of apparent turnover *T* was estimated by employing the equation of Diamond and May. The analysis confirmed that the longer the interval between successive surveys the lower the rate of 'apparent turnovers' *T*. However, because of the probable influence of a variety of factors, especially the very long gaps between some of the successive surveys, the results presented here should be regarded as preliminary rather than definitive. Hopefully the findings of this survey will encourage others to document (and report) the land bird species breeding on Clare Island, even during casual visits.

Although the rook is a regular visitor to the offshore islands of Ireland it does not breed on any of them. Chapter 8 is entitled 'Size matters: island area and breeding rooks *Corvus frugilegus* – an example of the Allee effect?'. Previously, it had been suggested that treeless habitats, such as occur on exposed Atlantic islands, might account for the absence of the rook from these locations. In this paper we set out to explain why the rook does not breed on Ireland's offshore islands. An analysis of the distribution of the rook on the offshore islands of England and Scotland suggests that a threshold area of ≥38,000ha is necessary for the species to establish a viable breeding population. The minimum area required to sustain a rookery exceeds by an order of magnitude the size of the largest offshore Irish island (Inishmore; 3,108ha). It appears that the Allee effect—where fitness increases with population density—appears to apply to the highly colonial breeding and roosting behaviour of the rook, and that if there is either an insufficiency or a reduction in its preferred grassland feeding habitat, its population will experience inverse density dependence and local extinction.

The future

On 24 July 1900, Richard Barrington (1900) in the introduction to his masterpiece 'The Migration of Birds as Observed at Irish Lighthouses and

Lightships' wrote that 'there are so many students nowadays, it becomes more difficult year by year to discover anything new, especially about our common English birds'. Of course, this was written before the discoveries of Gregor Mendel had gained wide currency, and the double helix structure of DNA, and the genetic code, had been revealed; therefore, the possibility of genetic studies on island populations of birds simply did not exist 120 years ago.

The situation has now changed dramatically, and the science of genomics has become firmly established in research into birds leading to new evolutionary, ecological and conservation-science perspectives (e.g. McMahon *et al.* (2012); Warren *et al.* (2015); Coates *et al.* (2018); Sangster (2014, 2018)).

Therefore, we suggest that a genomic based integrated taxonomic study should be carried out on some of the resident species on Clare Island, which may have been present since (and possibly before) its formation, including the rock pipit, wren and reed bunting *Emberiza schoeniclus* (Linnaeus). In addition, investigations into the genetics of the dunnock on Clare Island and indeed other offshore islands, may help to shed light on the conundrum posed by the findings of Paradis *et al.* (1998) and Wernham *et al.* (2002) on the one hand, and wide occupancy of the offshore islands of Britain and Ireland by this species (e.g. Balmer *et al.* 2013) on the other.

Given the large size of its breeding population (perhaps there are similar numbers on other offshore islands?) it would be of considerable ecological and conservation interest to conduct a detailed study into the breeding biology and general ecology of the wheatear on Clare Island. Hopefully such a study would include the use of tracking devices (e.g. Kays *et al.* 2015) to identify the course it takes to its winter quarters in Africa, the precise location of which is unknown. In a seabird study in the offshore waters of western Ireland (Mackey *et al.* 2004) wheatears were occasionally seen flying south in apparent ground effect (e.g. Finn *et al.* 2012). Do the Clare Island wheatears follow the same course as, for example, the whimbrel *Numenius phaeopus* (Linnaeus) (Alves *et al.* 2016) or does it take a different path or paths? (see also Bairlein *et al.* 2012).

The population dynamics of very few bird species have been researched in detail in Ireland and most studies are cross-sectional, that is, spanning two to three years study for example, for the duration of a PhD; only one passerine species, the dipper, is the subject of an ongoing longitudinal study now exceeding 30 consecutive years (e.g. Smiddy *et al.* 1995).

In addition to possible integrated taxonomic investigations, the rock pipit would appear to be a species worthy of a detailed population dynamics study, including especially its numerical abundance and demography. The rock pipit is situated in what has become a very perturbed breeding and feeding environment where a succession of major storms involving massive wave impacts on the coastline must have had negative effects on local populations, though this does not appear to have been investigated let alone quantified anywhere as yet. It would be important to establish how resilient its populations are as major storms are predicted to increase in the future.

The only systematic study of diurnal migration in the west of Ireland, carried out in Donegal, was that of Wallace *et al.* (2001). More investigations are necessary; Clare Island, in concert with Inishbofin, Tory Island and Inishmore, might be possible locations for such studies. The bulk of migration is nocturnal, and this can only be quantified by using radar (Eastwood 1967; see also Rosa *et al.* 2016 and references therein). Although originally prohibitively expensive, the development of relatively cheap 'Marine Radars' (used for example to monitor bird activity around windfarms) suggests that such a study could become possible in the future. At present, however, it is virtually impossible to identify the bird (or bat) species in the radar detected movements (Rosa *et al.* 2016). Bird vocalisations combined with radar may provide more detail about the taxa involved (Dr Gavin Fennessy, pers comm; see also Farnsworth *et al.* 2004).

REFERENCES

Alves, J.A., Dias, M.R., Mendez, V., Katrinaardottir, B. and Gunnarsson, T.G. 2016 Very rapid long distance sea crossing by a migratory bird. *Scientific Reports* **6**, 38154; doi:10.1038/srep38154.

Bairlein, F., Norris, D.R., Nagel, R., Bulte, M., Voight, C.C., Fox, J.W., Hussell, D.J. and Smaljohann, H. 2012 Cross-hemisphere migration of a 25g songbird. *Biological Letters* **8**, 505–7.

Balmer, D.E., Gillings, S., Caffrey, B.J., Swann, R.L., Downie, I.S. and Fuller, R.J. 2013 *Bird Atlas 2007–2011: the breeding and wintering birds of Britain and Ireland*. Thetford. BTO Books.

Barrington, R.M. 1900 *Migration of birds as observed at Irish lighthouses and lightships.* Dublin and London. Ponsonby and Porter.

Coates, D.J., Byrne, M. and Moritz, C. 2018 Genetic diversity and conservation units: dealing with the species – population continuum in the age of genomics. *Frontiers in Ecology and Evolution* **6**, Article 165.

Eastwood, E. 1967 *Radar ornithology*. London. Methuen.

Farnsworth, A., Gauthreaux, S.A. Jr and van Blaricom, D. 2004 A comparison of nocturnal call counts and reflectivity measurements on Doppler radar. *Journal of Avian Biology* **35**, 365–9.

Finn, J., Carlsson, J., Kelly, T. and Davenport, J. 2012 Avoidance of headwinds or exploitation of ground effect-why do birds fly low. *Journal of Field Ornithology* **83**, 192–202.

Kays, R., Crofoot, M.C., Jetz, W. and Wikelski, M. 2015 Terrestrial animal tracking as an eye on life and planet. *Science* **348**, doi: 10.1126 Science aaa2478–9.

Mackey, M., Ó Cadhla, O., Kelly, T.C., Aguilar de Soto, N. and Connolly, N. 2004 *Cetaceans and seabirds of Ireland's Atlantic margin*. Volume 1. Seabird distribution, density and abundance. Cork. Coastal and Marine Resource Centre.

McMahon, B.J., Johansson, M.P., Piertney, S.B., Buckley, K. & Hoglund, J. 2012 Genetic variation among endangered Irish red grouse (*Lagopus lagopus hibernicus*) populations: implications for conservation and management. *Conservation Genetics* **13**, 639–47.

Paradis, E., Baillie, S.R., Sutherland, W.J. and Gregory, R.D. 1998 Patterns of natal and breeding dispersal in birds. *Journal of Animal Ecology* **67**, 518–36.

Rosa, I.M.D., Marques, A.M., Palmunka, G.M., Costa, H., Mascarenhas, M., Fonesca, C. and Bernardino, J. 2016 Classification success of six machine learning algorithms in radar ornithology. *Ibis* **158**, 28–42.

Sangster, G. 2014 The application of species criteria in avian taxonomy and its implications for the debate of species concepts. *Biological Reviews* **89**, 199–214.

Sangster, G. 2018 Integrative taxonomy of birds: the nature and delimitation of species. In T. Tietze (ed) *Bird species, how they arise, modify and vanish.* 9–38. Switzerland. Springer.

Smiddy, P., O'Halloran, J., O'Mahony, B. and Taylor, A.J. 1995 The breeding biology of the dipper *Cinclus cinclus* in southwest Ireland *Bird Study* **42**, 76–81.

Wallace, D.I.M., McGeehan, A. and Allen, D. 2001 Autumn migration in westernmost Donegal. *British Birds* **94**, 103–20.

Warren, B.H., Simberloff, D., Ricklefs, R.E., Aguile, R., Condamine, F.L., Gravel, D., Morlon, H., Mouquet, N., Rosindell, J., Casquet, J., Cornault, J., Fernandez-Palacios, J.M., Hengl, T., Norder, S.J., Rijsdijk, K.F., Sanmartin, D., Triantis, K.A., Valente, L.M., Whiitaker, R.J., Gillespie, R.G., Emerson, B.C. and Thebaud, C. 2015 Islands as model systems in ecology and evolution: prospects fifty years after MacArthur-Wilson. *Ecology Letters* **18**, 200–17.

Wernham, C., Toms, M., Marchant, J., Clark, J., Siriwardina, G. and Baillie, S. 2002 *The migration atlas: movement of birds of Britain and Ireland.* London. Poyser.

Winters, P. 2006 *A Summer Bird Survey of Clare Island*. Unpublished MSc thesis. Dublin. Centre for the Environment, Trinity College.

Appendix 1: A list of land birds recorded on the 29 islands analysed in this study

(Each dot represents a breeding record for each species on each island, based on data from all published references available.)

	Clare Island Mayo	**Bills Rocks** Mayo	**Black Rock** Mayo	**Duvillaun More** Mayo	**Inishkeas** Mayo	**Inishmurray** Sligo	**Inishduff** Donegal
Area in ha	1598	4	5	63	327	85	3
Rock Dove	●			●	●	●	
Stock Dove							
Woodpigeon							
Collared Dove							
Cuckoo	●						
Great Spotted Woodpecker							
Skylark	●			●	●	●	
Sand Martin	●						
Swallow	●				●	●	
House Martin							
Meadow Pipit	●			●		●	
Rock Pipit	●	●	●	●	●	●	●
Grey Wagtail	●						
Pied Wagtail	●						
Dipper	●						
Wren	●				●	●	
Dunnock	●					●	
Robin	●					●	
Whinchat							
Stonechat	●						
Wheatear	●			●	●	●	
Ring Ouzel							
Blackbird	●					●	
Song Thrush	●					●	
Mistle Thrush							
Grasshopper Warbler	●						
Sedge Warbler	●					●	
Whitethroat	●					●	
Blackcap							
Garden Warbler							
Chiffchaff							
Willow Warbler	●						
Goldcrest							
Spotted Flycatcher							
Blue Tit							
Great Tit							
Coal Tit							
Long-tailed Tit							
Magpie	●						
Chough	●			●	●		
Jackdaw							
Rook							
Hooded Crow	●					●	
Raven	●			●	●	●	
Starling	●					●	
House Sparrow	●				●		
Tree Sparrow	●		●		●		
Chaffinch	●						
Greenfinch							
Goldfinch							
Siskin							
Linnet	●						
Twite	●						
Lesser Redpoll	●						
Yellowhammer	●						
Reed Bunting	●				●	●	
Corn Bunting	●						

	Inishbarnog Donegal	**Roaninish** Donegal	**Aranmore** Donegal	**Tory Island** Donegal	**Inishtrahull** Donegal	**Rathlin Island** Antrim	**Lambay Island** Dublin
Area in ha	5	10	1754	331	45	1437	250
Rock Dove				●	●	●	●
Stock Dove							●
Woodpigeon							●
Collared Dove			●			●	
Cuckoo			●			●	●
Great Spotted Woodpecker							●
Skylark		●	●	●		●	●
Sand Martin			●				
Swallow			●	●	●	●	●
House Martin						●	●
Meadow Pipit		●	●	●	●	●	●
Rock Pipit	●	●	●	●	●	●	●
Grey Wagtail						●	●
Pied Wagtail			●	●		●	●
Dipper							
Wren		●	●	●	●	●	●
Dunnock			●	●		●	●
Robin			●			●	●
Whinchat						●	
Stonechat			●		●	●	●
Wheatear		●	●	●	●	●	●
Ring Ouzel							●
Blackbird			●			●	●
Song Thrush			●			●	●
Mistle Thrush			●			●	●
Grasshopper Warbler						●	
Sedge Warbler			●	●		●	
Whitethroat			●			●	●
Blackcap						●	●
Garden Warbler							
Chiffchaff			●				●
Willow Warbler			●			●	●
Goldcrest			●			●	●
Spotted Flycatcher			●			●	●
Blue Tit						●	
Great Tit			●			●	
Coal Tit						●	
Long-tailed Tit							●
Magpie			●			●	●
Chough			●	●	●	●	
Jackdaw			●	●			●
Rook							●
Hooded Crow			●	●	●	●	●
Raven			●	●	●	●	●
Starling			●	●		●	●
House Sparrow			●	●		●	●
Tree Sparrow			●	●	●	●	
Chaffinch			●			●	●
Greenfinch			●			●	●
Goldfinch			●			●	●
Siskin						●	
Linnet		●	●	●		●	●
Twite				●		●	
Lesser Redpoll			●	●		●	
Yellowhammer			●	●		●	●
Reed Bunting			●			●	●
Corn Bunting			●	●		●	●

Appendix 1: A list of land birds recorded on the 29 islands analysed in this study (contd.)

(Each dot represents a breeding record for each species on each island, based on data from all published references available.)

	Great Saltee Wexford	Cape Clear Island Cork	Scariff Island Kerry	Deenish Island Kerry	Great Skellig Kerry	Great Blasket Kerry	Inishvickillaun Kerry
Area in ha	89	639	148	51	18	459	81
Rock Dove	●	●				●	●
Stock Dove		●					
Woodpigeon	●	●					
Collared Dove							
Cuckoo	●	●					
Great Spotted Woodpecker							
Skylark	●	●	●			●	●
Sand Martin						●	
Swallow	●	●				●	
House Martin	●	●				●	
Meadow Pipit	●	●	●			●	●
Rock Pipit	●	●	●	●	●	●	●
Grey Wagtail							
Pied Wagtail	●	●				●	
Dipper							
Wren	●	●	●	●	●	●	●
Dunnock	●	●				●	●
Robin	●	●				●	
Whinchat							
Stonechat	●	●	●	●		●	●
Wheatear	●	●	●	●	●	●	●
Ring Ouzel							
Blackbird	●	●		●			
Song Thrush	●	●				●	
Mistle Thrush		●					
Grasshopper Warbler	●						
Sedge Warbler	●	●					
Whitethroat	●	●					
Blackcap							
Garden Warbler	●						
Chiffchaff		●					
Willow Warbler		●					
Goldcrest							
Spotted Flycatcher	●						
Blue Tit		●					
Great Tit		●					
Coal Tit							
Long-tailed Tit							
Magpie		●					
Chough	●	●	●		●	●	●
Jackdaw	●	●					
Rook							
Hooded Crow	●	●	●	●	●	●	●
Raven	●	●	●	●	●	●	●
Starling	●	●					
House Sparrow		●				●	
Tree Sparrow							
Chaffinch		●					
Greenfinch		●					
Goldfinch	●	●					
Siskin							
Linnet	●	●					
Twite						●	
Lesser Redpoll							
Yellowhammer	●	●					
Reed Bunting	●	●					
Corn Bunting	●						

	Inishtearaght Kerry	**Beginish** Kerry	**Magharees** Kerry	**Inisheer** Galway	**Inishmaan** Galway	**Inishmore** Galway	**Inishbofin** Galway	**Inishshark** Galway
Area in ha	27	15	42	566	911	3090	936	235
Rock Dove			●			●	●	●
Stock Dove								
Woodpigeon				●	●	●		
Collared Dove				●		●	●	
Cuckoo				●	●	●	●	
Great Spotted Woodpecker								
Skylark			●	●	●	●	●	●
Sand Martin							●	
Swallow			●	●	●	●	●	●
House Martin				●				
Meadow Pipit			●	●	●	●	●	●
Rock Pipit	●	●	●	●	●	●	●	●
Grey Wagtail								
Pied Wagtail			●	●	●	●	●	
Dipper								
Wren	●	●	●	●	●	●	●	●
Dunnock	●			●	●	●	●	
Robin				●	●	●	●	●
Whinchat								
Stonechat				●	●	●	●	●
Wheatear	●	●	●	●	●	●	●	●
Ring Ouzel								
Blackbird				●	●	●	●	
Song Thrush					●	●	●	
Mistle Thrush					●			
Grasshopper Warbler								
Sedge Warbler					●		●	
Whitethroat						●	●	
Blackcap								
Garden Warbler								
Chiffchaff								
Willow Warbler								
Goldcrest								
Spotted Flycatcher								
Blue Tit								
Great Tit								
Coal Tit								
Long-tailed Tit								
Magpie					●	●	●	
Chough	●		●	●	●	●	●	●
Jackdaw			●	●	●	●		
Rook								
Hooded Crow				●	●	●	●	●
Raven	●					●	●	●
Starling				●	●	●	●	
House Sparrow				●	●	●	●	●
Tree Sparrow								
Chaffinch								
Greenfinch								
Goldfinch				●	●	●		
Siskin								
Linnet				●	●	●	●	●
Twite	●						●	●
Lesser Redpoll								
Yellowhammer							●	
Reed Bunting							●	
Corn Bunting							●	●

Appendix 2: Island data used in NMDS analysis

(Under Freshwater, Population past, Population present and Lighthouse the figures 0 and 1 represent negative and positive, respectively.)

Island	Area ha	Distance from mainland	Freshwater	Population past	Population present	Lighthouse	Latitude	Longitude
Clare	1598	5	1	1	1	0	53.804477	–9.9951075
Bills Rock	4	10	0	0	0	0	53.8725	–10.2067
Black Rock	5	13.5	0	0	0	1	54.067045	–10.319552
Duvillaun More	63	3.5	0	0	0	0	54.074599	–10.170325
Inishkeas	327	5	1	1	0	0	54.137633	–10.198298
Inishmurray	85	6.5	1	1	0	0	54.435556	–8.66
Inishduff	9	2.5	0	0	0	0	54.598485	–8.5460651
Inishbarnog	7	1.5	0	0	0	0	54.813889	–8.5597222
Roaninish	10	3.5	0	0	0	0	54.870779	–8.5354328
Aranmore	1754	3	1	1	1	0	54.987434	–8.5350585
Tory	331	12	1	1	1	0	55.265502	–8.2303706
Inishtrahull	45	7	0	0	0	1	55.4323	–7.2378879
Rathlin	1437	4.5	1	1	1	1	55.299936	–6.2184763
Lambay	250	4	0	1	1	0	53.490863	–6.0162586
Great Saltee	89	5.5	0	1	0	0	52.116667	–6.6166667
Cape Clear	639	6	1	1	1	0	51.438798	–9.4922606
Scariff	148	5.5	0	0	0	0	51.730152	–10.254971
Deenish	51	4	0	0	0	0	51.734736	–10.218937
Great Skellig	18	13	0	0	0	1	51.770662	–10.540524
Great Blasket	459	2	0	1	0	0	52.090824	–10.54106
Inishvickillaun	81	11.5	0	0	0	0	52.044347	–10.610469
Inishtearaght	27	14	0	0	0	1	52.074996	–10.656073
Beginish	15	2	0	0	0	0	52.115651	–10.507371
Magharees	42	2.5	0	0	0	0	52.326921	–10.022299
Inisheer	566	7.5	1	1	1	0	53.059442	–9.5281381
Inishmaan	911	13	1	1	1	0	53.085504	–9.5859717
Inishmore	3090	13	1	1	1	0	53.128872	–9.7196641
Inishbofin	936	6	1	1	1	0	53.620914	–10.215907
Inishshark	235	7.5	0	1	0	0	53.611667	–10.281667

BIBLIOGRAPHY

Abbott, I. 1980 Theories dealing with the ecology of landbirds on islands. *Advances in Ecological Research* **11**, 329–71.

Alerstam, T. 1990 *Bird Migration*. Cambridge. Cambridge University Press.

Alexander, S.M.D. 1954 The birds of the Blasket Islands, with special reference to the Great Blasket Island, Inishvickillaun and Illaunboy and some notes on the adjacent mainland. *Bird Study* **1**, 148–68.

Alldredge, M.W., Simons, T.R., and Pollack, K.H. 2007 Factors affecting aural detections of songbirds. *Ecological Applications* **17**, 948–55.

Alstrom, P., Rheindt, F.E., Zhang, R., Zhao, M., Wang, J., Zhu, X., Yin, C.G., Hao, Y., Ohlson, J., Jia, C., Prawiradialaga, D.M., Ericson, P.G.P., Lei, F. and Olsson, U. 2018 Complete species – level phylogeny of the leaf warbler (Aves: Phylloscopidae) radiation. *Molecular Phylogenetics and Evolution* **126**, 141–52.

Alves, J.A., Dias, M.R., Mendez, V., Katrinaardottir, B. and Gunnarsson, T.G. 2016 Very rapid long distance sea crossing by a migratory bird. *Scientific Reports* **6**, 38154; doi: 10.1038/srep38154.

Anonymous 2013 *Monthly Weather Bulletin, April 2013*. No. 324. Glasnevin. Met Éireann.

Anonymous 2013 *Monthly Weather Bulletin, May 2013*. No. 325. Glasnevin. Met Éireann.

Anonymous 2013 *Monthly Weather Bulletin, June 2013*. No. 326. Glasnevin. Met Éireann.

Anonymous 2015 Monthly Weather Bulletin, January 2015. No. 344 Glasnevin. Met Éireann.

Appelbaum, S. and Haney, A. 1981 Bird populations before and after wildfire in a Great Lakes pine forest. *Condor* **83**, 347–54.

Arlt, D., Forslund, P., Jeppsson, T. and Part, T. 2008 Habitat-specific population growth of a farmland bird. *PLoS One* **3**, e3006.

Avise, J.C. 1987 Mitochondrial DNA and the evolutionary genetics of higher animals. *Philosophical Transactions of the Royal Society London* **312B**, 325–42.

Avise, J.C. 1991 Ten unorthodox perspectives on evolution prompted by comparative population genetic findings on mitochondrial DNA. *Annual Review of Genetics* **25**, 45–69.

Avise, J.C. and Walker, D.E. 1998 Pleistocene phylogeographic effects on avian populations and the speciation process. *Proceedings of the Royal Society* **265B**, 457–63.

Ayala, F.J. and Fitch, W.M. 1997 Genetics and the origin of species: an introduction. *Proceedings of the National Academy of Sciences* **94**, 7691–7.

Bairlein, F., Norris, D.R., Nagel, R., Bulte, M., Voight, C.C., Fox, J.W., Hussell, D.J. and Smaljohann, H. 2012 Cross-hemisphere migration of a 25g songbird. *Biological Letters* **8**, 505–7.

Balmer, D., Gillings, S., Caffrey, B., Swann, B., Downie, I. and Fuller, R. 2013 *Bird Atlas 2007–2011: the breeding and wintering birds of Britain and Ireland*. Thetford. BTO Books.

Barlee, J. and Ruttledge, R.F. 1945 Notes on the present status of birds on Clare Island. *Irish Naturalists' Journal* **8**, 311–3.

Barrett-Hamilton, G.E.H. 1915 Mammalia. A biological survey of Clare Island in the county of Mayo, Ireland and adjoining district. *Proceedings of the Royal Irish Academy* **31,** Sections 1–3, Part 17.

Barrington, R.M. 1900 *The migration of birds as observed at Irish lighthouses and lightships*. Dublin and London. Ponsonby and Porter.

Barrowclough, G.R., Cracraft, J., Klica, J. and Zink, R.M. 2016 How many kinds of birds are there and why does it matter? *PLoS ONE* **11**, e0166307; doi: 10.1371/journal.pone.0166307.

Bauer, P., Thorpe, A. and Brunet, G. 2015 The quiet revolution of numerical weather prediction. *Nature* **525**, 47–55.

Bayley, S. and Caffrey, B. 2017 *Cape Clear Bird Observatory Annual Report 2016*. Report Number **25**. Wicklow. BirdWatch Ireland.

Bayley, S. and Caffrey, B. 2018 *Cape Clear Bird Observatory Annual Report 2017*. Report Number **26**. Wicklow. BirdWatch Ireland.

Beirne, B.P. 1952 *The origin and history of the British fauna*. London. Methuen.

Berec, L., Angulo, E. and Courchamp, F. 2007 Multiple Allee effects and population management. *Trends in Ecology and Evolution* **22**, 185–91.

Berrow, S. 1992 The diet of coastal breeding Ravens in Co. Cork. *Irish Birds* **4**, 555–8.

Berrow, S., Kelly, T.C. and Myers, A.A. 1991 Crows on estuaries: distribution and feeding behaviour of the Corvidae on four estuaries in southwest Ireland. *Irish Birds* **4**, 393–412.

Berrow, S., Kelly, T.C. and Myers, A.A. 1992 The diet of coastal breeding Hooded Crows *Corvus corone cornix*. *Ecography* **15**, 337–46.

Berrow, S., Kelly, T.C. and Myers, A.A. 1992 The mussel caching behaviour of Hooded Crows *Corvus corone cornix*. *Bird Study* **39**, 115–9.

Berry, R.J. 2009 *The natural history of islands*. London. Collins.

Betts, M. 1992 *Birds of Skokholm. Wales. Dyfed Wildlife Trust.*

Bibby, C.J., Burgess, N.D. and Hill, D.A. 1992 *Bird Census Techniques*. London. Academic Press.

Bibby, C.J., Burgess, N.D., Hill, D.A. and Mustoe, S. 2000 *Bird Census Techniques*. Second edition. London. Academic Press.

Bintanja, R., van de Wal, R.S. and Oerlaemans, J. 2005 Modelled atmospheric temperatures and global sea levels over the past million years. *Nature* **437**, 125–8.

Blackburn, T.M., Brown, V.K., Doube, B.M., Greenwood, J.J.D., Lawton, J.H. and Stork, N.E. 1993 The relationship between abundance and body size in natural animal assemblages. *Journal of Animal Ecology* **62**, 519–28.

Blackburn, T.M., Cassey, P. and Gaston, K.J. 2006 Variations on a theme: sources of heterogeneity in the form of the interspecific relationship between abundance and distribution. *Journal of Animal Ecology* **79**, 1426–39.

Blackburn, T.M., Gaston, K.J., Quinn, R.M., Arnold, H. and Gregory, R. 1997 Of mice and wrens: the relation between abundance and geographic range size in British mammals and birds. *Philosophical Transactions of the Royal Society of London* **352B**, 419–27.

Blondel, J. 1985 Breeding strategies of the Blue Tit and Coal Tit (*Parus*) in mainland and island Mediterranean habitats: a comparison. *Journal of Animal Ecology* **54**, 531–56.

Blondel, J. 2000 Evolution and ecology of birds on Islands: trends and prospects. *Vie et Milieu* **50**, 205–20.

Blondel, J., Perret, P., Maistre, M. and Dias, P.C. 1992 Do harlequin Mediterranean environments function as source or sink for Blue Tits (*Parus caeruleus* L.). *Landscape Ecology* **6**, 213–9.

Boddy, M. 1992 Timing of whitethroat Sylvia communis arrival, breeding and moult at a coastal site in Lincolnshire. *Ringing & Migration* **13**, 65–72.

Borton, D.G. 1987 The breeding passerines of Cape Clear Island 1986. Cape Clear Bird Observatory Report **19**, 50–61.

Boyle, D.P. 1997 A breeding survey of seabirds and other selected species on Cape Clear Island in 1995/1996. In Preston, K. (ed.) *Cape Clear Bird Observatory. Report Number Twenty Four*. 37–44. Dublin. Irish Wildbird Conservancy.

Bracken, F. and Bolger, T. 2006 Effects of set-aside management on birds breeding in lowland Ireland. *Agriculture, Ecosystems and Environment* **117**, 178–84.

Bracken, F. and Smiddy, P. 2012 Lowland bogs, Fens and Reed swamps. In R. Nairn and J. O'Halloran (eds) *Bird Habitats in Ireland*. 73–89. Cork. Collins Press.

Bracken, F., McMahon, B.J. and Whelan, J. 2008 Breeding bird populations of Irish peatlands. *Bird Study* **55**, 169–78.

Brooke, M. de L. 2002 Seabird systematics and distribution: a review of current knowledge. In E.A. Schreiber and J. Burger (eds) *Biology of marine birds*, 57–85. London, Boca Raton, New York and Washington. CRC Press.

Brooke, M.de L. and Davies, N.B. 1987 Recent changes in host usage by Cuckoos *Cuculus canorus* in Britain. *Journal of Animal Ecology* **56**, 873–83.

Brown, A. and Grice, P. 2005 *Birds in England*. London. Poyser.

Brown, C.R. and Rannala, B. 1995 Colony choice in birds: models based on temporally invariant site quality. *Behavioral Ecology and Sociobiology* **36**, 221–8.

Brown, D., Wilson, J., Douglas, D., Thompson, P., Foster, S., McCulloch, N., Phillips, J., Stroud, D., Whitehead, S., Crockford, N. and Sheldon, R. 2015 The Eurasian Curlew – the most pressing bird conservation priority in the UK? *British Birds* **108**, 660–8.

Buchanan, G.M., Grant, M.C., Sanderson, R.A. and Pearce-Higgins, J.W. 2006 The contribution of invertebrate taxa to moorland bird diets and potential implications for land–use management. *Ibis* **148**, 615–28.

Buckland, S.J., Anderson, D.R., Burnham, K.P., Laake, J-L., Bochers, D.L. and Thomas, L. 2001 *Introduction to Distance Sampling: estimating abundance of biological populations*. Oxford. Oxford University Press.

Buckley, N.J. 1990 Diet and feeding ecology of great black-backed gulls (Larus marinus) at a southern Irish breeding colony. *Journal of Zoology (London)* **222**, 363–73.

Buckley, N.J. and Kelly, T.C. 1994 Breeding biology of great black-backed gulls Larus marinus at a declining colony: Cape Clear Island, Co. Cork. *Irish Naturalists' Journal* **24**, 388–92.

Bull, J., Wanless, S., Elston, D.A., Daunt, F., Lewis, S. and Harris, M.P. 2004 Local scale variability in the diet of black-legged kittiwakes Rissa tridactyla. *Ardea* **92**, 43–52.

Burnham, K.P. and Anderson, D.R. 1984 Tests of compensatory versus additive mortality in mallards. *Ecology* **65**, 105–12.

Burnham, K.P. and Anderson, D.R. 1984 The need for distance data in transect counts. *Journal of Wildlife Management* **48**, 1248–54.

Butlin, R.K., Galindo, J. and Grahame, J.W. 2008 Sympatric, parapatric or allopatric: the most important way to classify speciation. *Philosophical Transactions of the Royal Society* **363B**, 2997–3007.

BWP Birds of the Western Palaearctic Interactive. BWPi Version 2.

Cabot, D., Cabot, R. and Viney, M. 2020 The breeding seabirds and other birds on the Bills Rocks, County Mayo. *Irish Birds* **42**, 55–62.

Cabot, D.B. 1962 An ornithological expedition to Inishmurray, Co. Sligo. *Irish Naturalists Journal* **14**, 59–61.

Cabot, D.B. 1963 The breeding birds of the Inishkea Islands, Co. Mayo. *Irish Naturalists' Journal* **14**, 113–5.

Calladine, J., Pakeman, R., Humphreys, E., Huband, S. and Fuller, R.J. 2014 Changes in breeding wader assemblages, vegetation and land use within machair environments over three decades. *Bird Study* **61**, 287–300.

Canestrelli, C., Bisconti, R. and Carere, C. 2015 Bolder takes all? Behavioural dimension to biogeography. *Trends in Ecology and Evolution* **31**, 35–43.

Castle, M.E. 1977 Rookeries in Scotland – 1975. *Scottish Birds* **9**, 327–34.

Chivers, L.S. 2008 Breeding seabirds on Rathlin Island, County Antrim – a cause for concern. *Irish Birds* **8**, 359–64.

Chivers, L.S., Lundy, M.G. and Reid, N. 2012b Stable breeding despite variable feeding in two sympatric auks (Alcidae) species. *Bird Study* **59**, 67–73.

Chivers, L.S., Lundy, M.G., Colhoun, K., Newton, S.F. and Reid, N. 2012a Diet of black-legged kittiwakes (Rissa tridactyla) feeding chicks at two Irish colonies highlights the importance of clupeids. *Bird Study* **59**, 363–7.

Clancy, C.F., O'Callaghan, M.J.A. and Kelly, T.C. 2006 A multiscale problem arising in a model of avian flu virus in a seabird colony. *Journal of Physics* (conference series) **55**, 45–54.

Claramunt, S., Derryberry, E.P., Remsen J.V. and Brumfield, R.T. 2012 High dispersal ability inhibits speciation in a continental radiation of passerine birds. *Proceedings of the Royal Society London* **279B**, 1567–74.

Clegg, S.M. and Owens, I.P.F. 2002 The 'island rule' in birds: medium body size and its ecological explanation. *Proceedings of the Royal Society, London* **269B**, 1359–65.

Clobert, J., Danchin, E., Dhondt, A.A. and Nicols, J.D. 2001 *Dispersal*. Oxford. Oxford University Press.

Clobert, J., Le Gaillard, J.F., Cote, J., Meylan, S. and Massot, M. 2009 Informed dispersal, heterogeneity in animal dispersal syndromes and the dynamics of spatially structured populations. *Ecology Letters* **12**, 197–209.

Coates, D.J., Byrne, M. and Moritz, C. 2018 Genetic diversity and conservation units: dealing with the species – population continuum in the age of genomics. *Frontiers in Ecology and Evolution* **6**, Article 165.

Colhoun, K. and Cummins, S. 2013 Birds of Conservation Concern in Ireland. *Irish Birds* **9**, 523–44.

Colhoun, K., Mawhinney, K. and Peach, W.J. 2015 Population estimates and changes in abundance of breeding waders in Northern Ireland up to 2013. *Bird Study* **62**, 394–403.

Collins, T.C. 1999 The Clare Island survey of 1909–1911: participants, papers and progress. In C. Mac Carthaigh and K. Whelan (eds) *New Survey of Clare Island. Volume **1**: History and cultural landscape*, 1–40. Dublin. Royal Irish Academy.

Connor, E.F. and McCoy, E.D. 1979 The statistics and biology of the species area relationship. *American Naturalist* **13**, 791–833.

Coombes, D. 2003 Clare Island. Birds noted on the island during a 3-day visit 29/4 to 1/5/2000. Unpublished report.

Coombs, F. 1978 *The crows*. London. Batsford.

Corlett, R.T. 2015 The Anthropocene concept in ecology and conservation. *Trends in Ecology and Evolution* **30**, 36–41.

Coulson, J.C. 2002 Colonial breeding in seabirds. In E.A. Schreiber and J. Burger (eds) *Biology of marine birds*, 87–113. London, Boca Raton, New York and Washington. CRC Press.

Courchamp, F., Clutton-Brock, T. and Grenfell, B. 1999 Inverse density dependence and the Allee effect. *Trends in Ecology and Evolution* **14**, 405–10.

Coxon, P. 2001 The Quaternary History of Clare Island. In J.R. Graham (ed.) *New Survey of Clare Island. Volume **2**: Geology*, 87–112. Dublin. Royal Irish Academy.

Coxon, P. 2008 Landscapes and environments in the late glacial-interglacial transition: a time of amazingly rapid change in Ireland. In J.L. Davenport, D.P. Sleeman and P.C. Woodman (eds) *Mind the gap: postglacial colonization of Ireland*. 45–61. Belfast. Irish Naturalists' Journal.

Coxon, P., Corcoran., R., Gibson, P. and McCarron, S. 2013 A Holocene pollen diagram from Lough Avullin, Clare Island, Western Ireland. In D. Synnott (ed.) *New Survey of Clare Island. Volume **7**: Plants and fungi*, 1–26. Dublin. Royal Irish Academy.

Coyne, J.A. and Orr, A. 2004 *Speciation*. Massachusetts, Sinaur Associates.

Craig, R.J., Suh, A., Wang, M. and Ellegren, H. 2018 Natural selection beyond genes: identification and analyses of evolutionary conserved elements in the genome of the collared flycatcher (*Ficedula albicollis*). *Molecular Ecology* **27**, 476–92.

Cramp, S. and Perrins, C.M. 1994 *The birds of the Western Palearctic*. Volume 6. Oxford. Oxford University Press.

Cramp, S. and Simmons, K.E.L. (eds) 1983 *Handbook of birds of Europe, The Middle East and North Africa*. Vol. 3. Waders to Gulls. Oxford. Oxford University Press.

Cramp, S., Bourne, W.R.P. and Saunders, D. 1974 *The seabirds of Britain and Ireland*. London. Collins.

Crème, G. 1997 Aspects of the biology of an expanding population of lesser black-backed gulls *Larus fuscus* in southwest Ireland. Unpublished Ph.D. thesis. Cork. University College.

Crowe, O., Coombes, R.H., Lysaght, L., O'Brien, C., Choudhury, K.R., Walsh, A.J., Wilson, J.H. and O'Halloran, J. 2010 Population trends of widespread breeding birds in the Republic of Ireland 1998–2008. *Bird Study* **57**, 267–80.

Crowe, O., Musgrove, A. and O'Halloran, J. 2014 Generating population estimates for common and widespread breeding birds in Ireland. *Bird Study* **61**, 82–90.

Croxall, J.P. (ed.) 1987 *Seabirds–feeding ecology and role in marine ecosystems*. Cambridge. Cambridge University Press.

Cummins, S. and O'Halloran, J. 2003 The breeding biology of the stonechat Saxicola torquata in southwest Ireland. *Irish Birds* **7**, 177–86.

Cummins, S., Bleasdale, A., Douglas, C., Newton, S., O'Halloran, J. and Wilson, H.J. 2010 The status of the red grouse in Ireland and the effects of land use, habitat and habitat quality on their distribution. Results from the National Red Grouse Survey 2007–2008. Irish Wildlife Manuals No. 50. Dublin. National Parks and Wildlife Service.

Cummins, S., Bleasdale, A., Douglas, C., Newton, S., O'Halloran, J. and Wilson, H.J. 2015 Densities and population estimates of red grouse *Lagopus lagopus scotica* in Ireland based on the 2006–2008 national survey. *Irish Birds* **10**, 197–210.

Cummins, S., O'Connell, P. and Mee, A. 2012 Upland heath and blanket bog. In R. Nairn and J. O'Halloran (eds) *Bird Habitats in Ireland*. 90–104. Cork. Collins Press.

Cury, P.M., Boyd, I.L., Bonhommeau, S., Anker-Nilssen, T., Crawford, R.M., Furness, R.W., Mills, J.A., Murphy, E.J., Osterblom, H., Paleczny, S., Platt, J.F., Roux, J-P., Shannon, L. and Sydeman, W.J. 2011 Global seabird response to forage fish depletion – one third for the birds. *Science* **334**, 1703–6.

Cussen, R. and Myers, A. 2002 Dog whelk Nucella lapillus (Linnaeus), a significant prey item in the diet of herring gulls on Clare Island. In A. Myers (ed.) *New Survey of Clare Island, Volume 3: Marine intertidal ecology*, 151–7. Dublin. Royal Irish Academy.

Cussen, R., Kelly, T.C., Hartnett, M. and Walsh, P.M. 1999 Counts of breeding seabirds on Clare Island, Co. Mayo, 1999. Dublin. Report to Royal Irish Academy.

Cussen, R., Kelly, T.C., Hartnett, M. and Walsh, P.M. 1999 Counts of breeding seabirds on Clare Island, Co. Mayo, 1999. Dublin. Report to Royal Irish Academy for the New Survey of Clare Island.

D'Arcy, G. 1992 Fauna observed on Clare Island, Co. Mayo June 14–18 1990. *Irish Naturalists' Journal* **24**, 33–4.

Dai, L., Vorselan, D., Korolev, K.S. and Gore, J. 2012 Generic indicators for loss of resilience before a tipping point leading to population collapse. *Science* **336**, 1175–7.

Danchin, E. and Wagner, R.H. 1997 The evolution of coloniality: the emergence of new perspectives. *Trends in Ecology and Evolution* **12**, 342–7.

Danchin, E., Giraldeau, L-A., Valone, T.J. and Wagner, R.H. 2004 Public information: from noisy neighbours to cultural evolution. *Science* **305**, 487–91.

Darwin, C. 1845 *The Voyage of the Beagle*. London. John Murray.

Darwin, C. 1859 *On the origin of species by means of natural selection, or the preservation of favoured races in the struggle for life*. London. John Murray.

Davenport, J., O'Callaghan, M.J.A., Davenport, J.K. and Kelly, T.C. 2014 Mussel dropping by Carrion and Hooded Crows: biomechanical and energetic considerations. *Journal of Field Ornithology* **85**, 196–205.

Davenport, J.L., Sleeman, D.P. and Woodman, P.C. (eds) 2008 *Mind the gap: postglacial colonization of Ireland*. Special Supplement 2008. Belfast. Irish Naturalists' Journal.

Davies, N. 2015 *Cuckoo: cheating by nature*. London. Bloomsbury.

de Courcey Ireland, J. 1981 *Ireland's Sea Fisheries – a history*. Dublin. The Glendale Press.

Delany, S., Scott, D., Dodman, T. and Stroud, D. (eds) 2009 *An Atlas of Wader Populations in Africa and Western Eurasia*. Wageningen. Wetlands International.

Dempsey, E. and O'Clery, M. 2007 *Finding birds in Ireland: The complete guide*. Dublin. Gill and Macmillan.

Diamond, J. and May, R.M. 1 977 Species turnover rates on islands: dependence on census interval. *Science* **197**, 266–70.

Dolton, C.S. and Brooke, M.de L. 1999 Changes in the biomass of birds breeding in Great Britain 1968–88. *Bird Study* **46**, 274–8.

Donaghy, A., Prosser, C. and Lynch, A. 2014 Factors affecting breeding success of waders at the Shannon Callows. *Irish Birds* **10**, 37–46.

Donald, P.F., Wilson, J.D. and Shepherd, M. 1994 The decline of the corn bunting. *British Birds* **87**, 106–32.

Doyle, G.J. and Foss, P.J. 1986 A resurvey of the Clare Island flora. *Irish Naturalists' Journal* **22**, 85–9.

Dunning, J.B. (ed) 2008 *CRC Handbook of avian body masses*. Second edition. London. CRC Press.

Durant, J.M., Hjermann, D.O., Frederiksen, M., Charrassin, J.B., Le Maho., Y., Sabarros, P.S., Crawford, R.J.M. and Stenseth, N.C. 2009 Pros and cons of using seabirds as ecological indicators. *Climate Research* **39**, 115–29.

Dwyer, D.J. 1963 Farming an Atlantic Outpost: Clare Island, Co. Mayo. *Geography* **48**, 255–67.

Dymond, J.N. 1980 *The birds of Lundy*. Plymouth. Norwood, Roose and Co. Ltd.

Dymond, J.N. 1991 *The birds of Fair Isle*. Edinburgh. Ritchie.

Eastwood, E. 1967 *Radar ornithology*. London. Methuen.

Ellegren, H. 2013 The evolutionary genomics of birds. Annual Review of Ecology. *Evolution and Systematics* **44**, 239–59.

Elliot, E.C. and Cornell, J. 2012 Dispersal polymorphism and the speed of biological invasions. *PLoS One* **7**, 00404496.

Elliott, R.E. 1989 *The birds of Islay*. London. Christopher Helm.

Evans, K.L., James, N.A. and Gaston, K.J. 2006 Abundance, species richness and energy availability in the North American avifauna. *Global Ecology and Biogeography* **15**, 372–85.

Evans, K.L., Leech, D.I., Crick, H.Q.P., Greenwood, J.J.D. and Gaston, K.J. 2009 Latitudinal and seasonal patterns in clutch size of some single brooded British birds. *Bird Study* **56**, 75–85.

Ewins, P.J., Dymond, J.N. and Marquiss, M. 1986 The distribution, breeding and diet of Ravens *Corvus corax* on Shetland. *Bird Study* **33**, 110–6.

Fahrig, L. 2001 How much habitat is enough? *Biological Conservation* **100**, 65–74.

Fahy, K. 2012 Irish Rare Bird Report 2011. *Irish Birds* **9**, 447–93.

Fairley, J.S. 1981 *Irish whales and whaling*. Belfast. Blackstaff Press.

Farnsworth, A., Gauthreaux, S.A. Jr and van Blaricom, D. 2004 A comparison of nocturnal call counts and reflectivity measurements on Doppler radar. *Journal of Avian Biology* **35**, 365–9.

Feehan, J. 2019 *Clare Island*. Dublin. Royal Irish Academy.

Fennessy, G. 2001 Aspects of the ecology of the Robin *Erithacus rubecula* L. Ph.D. thesis. Cork. National University of Ireland.

Ferguson-Lees, J., Castell, R. and Leech, D. 2011 *A field guide to monitoring nests*. Thetford. British Trust for Ornithology.

Fernandez-Palacios, J.M. 2016 Shaped by sea level shifts. *Nature* **532**, 42–3.

Fernandez-Palacios, J.M., Rijsdijk, K.F., Norder, S.J., Otto, R., de Nascimento, L., Fernandez-Lugo, S., Tjorve, E. and Whiitaker, R.J. 2015 Towards a glacial-sensitive model of island biogeography. *Global Ecology and Biogeography* **25**, 817–30.

Fǐser, C., Robinson, C.T. and Malard, F. 2018 Cryptic species as a window into the paradigm shift of the species concept. *Molecular Ecology* **27**, 613–35.

Finn, J., Carlsson, J., Kelly, T. and Davenport, J. 2012 Avoidance of headwinds or exploitation of ground

effect-why do birds fly low. *Journal of Field Ornithology* **83**, 192–202.
Finney, K. and Donaghy, A. 2014 The effect of management measures on breeding wader populations in the Shannon Callows. *Irish Birds* **10**, 1–10.
Fisher, J. 1952 *The fulmar*. New Naturalist. London. Collins.
Fisher, J. 1966 The fulmar population of Britain and Ireland, 1959. *Bird Study* **13**, 5–76.
Forrester, R., Andrews, I., McInerney, C., Murray, R., McGowan, R., Zonfrillo, B., Betts, M., Jardine, D. and Grundy, D. (eds) 2007 *The birds of Scotland*. Aberlady. Scottish Ornithologists' Club.
Förschler, M.I. and Geiter, O. 2010 Morphological shifts in House Sparrows *Passer domesticus* in Helgoland. *Ornis Fennica* **87**, 93–8.
Forsyth, I. and Allen, P. 1997 *Report to the Royal Irish Academy on the breeding birds of Clare Island, Co. Mayo, 1996 and 1997*. 9 pages. Unpublished report to Royal Irish Academy.
Frederiksen, M.F., Mavor, R.A. and Wanless, S. 2007 Seabirds as environmental indicators: the advantages of combining data sets. *Marine Ecology Progress Series* **352**, 205–11.
Frederiksen, M.F., Wright, P.J., Harris, M., Mavor, R.A., Heubeck, M. and Wanless, S. 2005 Regional patterns of kittiwake Rissa tridactyla breeding success are related to variability in sandeel recruitment. *Marine Ecology Progress Series* **300**, 201–11.
Freeland, J.R., Andersson, S., Allen, D. and Looney, D. 2007 Museum samples provide novel insights into the taxonomy and genetic diversity of Irish red grouse. *Conservation Genetics* **8**, 695–703.
Fretwell, S.D. and Lucas, H.L. Jr. 1970 On territorial behaviour and the factors influencing habitat distribution in birds. 1. Theoretical development. *Acta Biotheoretica* **19**, 1–36.
Fuller, R.J. (ed) 2012 *Birds and habitat. Relationships in changing landscapes.* Cambridge. Cambridge University Press.
Fuller, R.J., Gaston, K.J. and Quine, C.P. 2007 Living on the edge: British and Irish woodland birds in a European context. *Ibis* **149** 56–63.
Fulton, D. 2010 The breeding population of Northern Wheatears at Clee Hill, Shropshire, 1998–2009. *British Birds* **103**, 223–8.
Furness, R.W. 1978 Energy requirements of seabird communities: a bioenergetics model. *Journal of Animal Ecology* **47**, 39–53.
Furness, R.W. 2007 Responses of seabirds to depletion of food fish stocks. *Journal of Ornithology* **148**, 247–52.
Gaston, K.J. 1994 *Rarity*. London. Chapman and Hall.
Gaston, K.J., Blackburn, T.M., Greenwood, J.J.D., Gregory, R.D., Quinn, R.M. and Lawton, J.H. 2000 Abundance occupancy relationships. *Journal of Applied Ecology* **37** (Supplement 1), 39–59.
Gibbons, D.W., Reid, J.B. and Chapman, R.A. (eds) 1993 *The new atlas of breeding birds in Britain and Ireland: 1988–1991*. London. Poyser.
Gill, F.B. 2014 Species taxonomy of birds: which null hypothesis? *The Auk* **131**, 150–61.
Gill, F.B. and Donsker, D. (eds) 2018 IOC world bird list. Version 8.2. www.worldbirdnames.org.
Gillings, S. and Newson, S.E. 2007 Breeding population estimate for Northern Wheatear in Britain. *British Birds* **100**, 179–84.
Glue, D. and Morgan, R. 1972 Cuckoo hosts in British habitats. *Bird Study* **19**, 187–92.
Goldstein, D.L. 2002 Water and salt balance in seabirds. In E.A. Schreiber and J. Burger (eds) *Biology of marine birds*, 467–83. London, Boca Raton, New York and Washington. CRC Press.
Gosling, P. 2007 The human settlement history of Clare Island. In P. Gosling, C. Manning and J. Waddell (eds) *New Survey of Clare Island. Volume 5: Archaeology*, 29–68. Dublin. Royal Irish Academy.
Gottschalk, T.K. and Huettmann, F. 2010 Comparison of distance sampling and territory mapping methods for birds in four different habitats. *Journal of Ornithology* **152**, 421–9.
Graham, C.T., Wilson, M.W., Gittings, T., Kelly, T.C., Irwin, S., Quinn, J.L. and O'Halloran, J. 2015 Implications of afforestation for bird communities: the importance of preceding land-use type. *Biodiversity Conservation.* DOI 10.1007/s10531–015–0987–4.
Grant, M.C., Orsman, C., Easton, J., Lodge, C., Smith, C., Thompson, G., Rodwell, S. and Moore, N. 1999 Breeding success and causes of breeding failure of Curlew *Numenius arquata* in Northern Ireland. *Journal of Applied Ecology* **36**, 59–74.
Gray, N., Thomas, G., Trewby, M. and Newton, S.F. 2003 The status and distribution of Choughs Pyrrhocorax pyrrhocorax in the Republic of Ireland 2002–2003. *Irish Birds* **7**, 147–57.
Greenwood, P. 1980 Mating systems, philopatry and dispersal in birds and mammals. *Animal Behaviour* **28**, 1140–62.
Griffin, L.R. 1999 Colonization patterns of rook *Corvus frugilegus* colonies: implications for survey strategies. *Bird Study* **46**, 170–3.
Griffin, L.R. and Thomas, C.J. 2000 The spatial distribution and size of rook (*Corvus frugilegus*) breeding colonies is affected by both the distribution of foraging habitat and by intercolony competition. *Proceedings of the Royal Society* **267B**, 1463–7.
Harris, M.P. and Wanless, S. 1996 Differential responses of guillemot Uria aalge and shag Phalacrocorax aristotelis to a late winter wreck. *Bird Study* **43**, 220–30.
Harris, M.P., Heubeck, M., Newell, M.A. and Wanless, S. 2015a The need for year-specific correction factors (k values) when converting counts of individual common guillemots Uria aalge to breeding pairs. *Bird Study* **62**, 276–9.
Harris, M.P., Newell, M.A. and Wanless, S. 2015b The use of k values to convert counts of individual razorbills Alca torda to breeding pairs. *Seabird* **28**, 30–6.
Hart, H.C. 1883 The birds of Lambay Island, Co. Dublin. *Zoologist* (3) **7**, 155–64, 225–6.
Heaney, L.R. 2000 Dynamic disequilibrium: a long-term large scale perspective on the equilibrium model of

island biogeography. *Global Ecology and Biogeography* **9**, 59–74.
Heath, M.F., Borrgreve, C. and Peet, N. 2000 *European bird populations: estimates and trends*. Cambridge. Birdlife International.
Heery, S. 2018 Charles Joseph Patten: professor and Irish ornithologist. *Irish Birds* **11**, 101–5.
Helmus, M.R., Mahler, L. and Losos, J.B. 2014 Island biogeography of the Anthropocene. *Nature* **513**, 543–6.
Henderson, I., Holt, C. and Vickery, J. 2007 National and regional patterns of habitat association with foraging Barn Swallows *Hirundo rustica* in the UK. *Bird Study* **54**, 371–7.
Henderson, I.G., Wilson, A.M., Steele, D. and Vickery, J.A. 2002 Population estimates, trends and habitat associations of breeding Lapwings *Vanellus vanellus*, Curlew *Numenius arquata* and Snipe *Gallinago gallinago* in Northern Ireland in 1999. *Bird Study* **49**, 17–25.
Heraughty, P. 1982 *Inishmurray ancient monastic island*. Dublin. O'Brien Press.
Hewell-Smith, H. 2004 *The Scottish Islands*. Edinburgh. Canongate.
Hillis, J.P. 2010 Rare Irish breeding birds, 2009. The annual report of the Irish Rare Breeding Birds Panel (IRBBP). *Irish Birds* **9**, 67–76.
Hobbs, J. 2018 A list of Irish birds. Version.5.2, July 2018. www.southdublinbirds.com.
Holland, J.M., Hutchison, M.A.S., South, B. and Aebischer, N.J. 2006 A review of invertebrate and seed bearing plants as food of farmland birds in Europe. *Annals of Applied Biology* **148**, 49–71.
Hounsome, M.V. and Rear, D. 1966 The bird population of the Great Blasket Island, Illaunboy and Beginish, Co. Kerry. *Irish Naturalists' Journal* **15**, 169–75.
Hudson, P.J. and Cattadori, I.M. 1999 The Moran Effect: a cause of population synchrony. *Trends in Ecology and Evolution* **14**, 1–2.
Hutchinson, C.D. (ed.) 1975 *The birds of Dublin and Wicklow*. Dublin. Irish Wildbird Conservancy.
Hutchinson, C.D. 1989 *Birds in Ireland*. Calton. Poyser.
Hutchinson, C.D. 1997 Bird study in Ireland. In J.W. Foster (ed.) *Nature in Ireland – a scientific and cultural history*, 262–82. Dublin. Lilliput Press.
Hutchinson, C.D. and Ruttledge, R.F. 1978 The birds of Inishbofin and Inishshark, Co. Galway. *Irish Birds* **1**, 211–24.
Hutto, R.L. 1990 Measuring the availability of food resources. *Studies in Avian Biology* **13**, 20–8.
Ivar do Sol, J.A. and Costa, M.F. 2014 The present and future of microplastic pollution of the marine environment. *Environmental Pollution* **185**, 352–64.
Jackson, D.B. 2001 Experimental removal of introduced hedgehogs improves wader nest success in the Western isles, Scotland. *Journal of Applied Ecology* **38**, 802–12.
Jackson, D.B. and Green, R.E. 2000 The importance of the introduced hedgehog (*Erinaceus europeus*) as a predator of the eggs of waders (Charadrii) on Machair, on South Uist, Scotland. *Biological Conservation* **93**, 333–48.
Jackson, D.B., Fuller, R.J. and Campbell, S.T. 2004 Long-term population changes among shorebirds in the Outer Hebrides, Scotland, in relation to introduced hedgehogs (*Erinaceus europeus*). *Biological Conservation* **117**, 151–66.
Jessopp, M. and Kelly, T.C. 2012 Open sea. In R. Nairn and J. O'Halloran (eds) *Bird habitats in Ireland*, 181–95. Cork. Collins Press.
JNCC 1999 Seabird 2000 – census instructions. Aberdeen. Joint Nature Conservation Committee.
JNCC 2015 Seabird population trends and causes of change: 1986–2014 report (http://www.jncc.defra.gov.uk/ page-3201). Aberdeen. Joint Nature Conservation Committee. Updated October 2015. Accessed 21 July 2016.
Jurasinski, G., Retzer, V. and Beirerkuhnlein, C. 2009 Inventory, differentiation and proportional diversity: a consistent terminology for quantifying species diversity. *Oecologia* **159**, 15–26.
Kapusta, A. and Suh, A. 2017 Evolution of bird genomes – a transposon's eye view. *Annals of the New York Academy of Sciences* **1389**, 164–8.
Karr, J.R. 1968 Habitat and avian diversity on strip-mined land in east-central Illinois. *Condor* **70**, 348–57.
Kays, R., Crofoot, M.C., Jetz, W. and Wikelski, M. 2015 Terrestrial animal tracking as an eye on life and planet. *Science* **348**, doi: 10.1126 Science aaa2478–9.
Kelleher, K.M. and O'Halloran, J. 2006 Breeding biology of the Song Thrush *Turdus philomelos* in an island population. *Bird Study* **53**, 142–55.
Kelly, M., Oxbrough, A. and Kelly, T.C. 2012 Some observations on the diet of adult common gulls (*Larus canus*) at a coastal breeding colony in County Kerry. *Irish Birds* **9**: 357–60.
Kelly, T. 2008 The origin of the avifauna of Ireland. In J.L. Davenport, D.P. Sleeman and P.C. Woodman. (eds) *Mind the gap: post glacial colonisation of Ireland*. 97–107. Belfast. Irish Naturalists' Journal Special Supplement.
Kelly, T.C. 2000 and 2013 Unpublished survey data.
Kelly, T.C. 2013 Unpublished data, Clare Island, Co. Mayo. Unpublished report.
Kelly, T.C., Smiddy, P. and Graham, C.T. 2014 Origins of the Irish avifauna – a perspective from Island Biography Theory. In D.P. Sleeman, J. Carlsson and J.E.L. Carlsson (eds) *Mind the gap II. New insights into the Irish postglacial*. 44–52. Belfast. Irish Naturalists' Journal.
Kennedy, P.G., Ruttledge, R.F. and Scroope, C.F. 1954 *The birds of Ireland*. London and Edinburgh. Oliver and Boyd.
Kennedy, P.G. 1957 Birds. In J. Meenan and D.A. Webb (eds) *A View of Ireland*. 62–8. British Association for the Advancement of Science. Dublin. Hely's Limited.
Kirwan, G., Porter, R. and Scott, D. 2015 Chronicle of an extinction? A review of Slender-billed Curlew records in the Middle East. *British Birds* **108**, 669–82.
Koenig, W.D. 2002 Global patterns of environmental synchrony and the Moran Effect. *Ecography* **25**, 283–8.
Koenig, W.D., Von Vuren, D. and Hooge, P.N. 1996 Detectability, philopatry and the distribution of

dispersal distances in vertebrates. *Trends in Ecology and Evolution* **11**, 514–7.

Lack, P.C. 1986 *The atlas of wintering birds in Britain and Ireland*. Calton. Poyser.

Lande, R. 1987 Extinction thresholds in demographic models of terrestrial populations. *American Naturalist* **130**, 624–35.

Lawton, J.H. 1989 What is the relationship between population density and size in animals? *Oikos* **55**, 429–34.

Lloyd, C.S. 1983 The status of landbirds breeding on Great Saltee, Co. Wexford. *Irish Naturalists' Journal* **21**, 97–103.

Lloyd, C.S. 1984 The birds of Clare Island, Co. Mayo in June 1982. *Irish Naturalists' Journal* **21**, 212–6.

Lloyd, C.S., Tasker, M.L. and Partridge, K. 1991 *The status of seabirds in Britain and Ireland*. Calton. Poyser.

Lomolino, M.V. 2005 Body size evolution in insular vertebrates: generality of the island rule. *Journal of Biogeography* **32**, 1683–99.

Losos, J.B. and Ricklefs, R. (eds) 2010 *The theory of island biogeography revisited*. Princeton and Oxford. Princeton University Press.

Losos, J.B. and Ricklefs, R. 2009 Adaptation and diversification on islands. *Nature* **457**, 830–6.

Lovegrove, R., Williams, G. and Williams, I. 1994 *Birds in Wales*. London. Poyser.

Lusher, A.L., Burke, A., O'Connor, I. and Officer, M. 2014 Microplastic pollution in the Northeast Atlantic: validated and opportunistic sampling. *Marine Pollution Bulletin* **88**, 325–33.

Luther, D. and Greenberg, R. 2011 The island syndrome in coastal wetland ecosystems: convergent evolution in large bills in mangrove passerines. *The Auk* **28**, 201–4.

Lysaght, L. 1989 Breeding bird populations of farmland in mid-west Ireland. *Bird Study* **36**, 91–8.

Lysaght, L. 2002 *An atlas of breeding birds of the Burren and the Aran Islands*. Dublin. BirdWatch Ireland.

Lysaght, L. and Marnell, F. (eds) 2016 *Atlas of mammals in Ireland 2010–2015*. Waterford. National Biodiversity Data Centre.

Mac Cárthaigh, C. 1999 Clare Island folklife. In C. Mac Cárthaigh and K. Whelan (eds) *New Survey of Clare Island. Volume 1: History and cultural landscape*, 41–72. Dublin. Royal Irish Academy.

MacArthur, R.H. and Wilson, E.O. 1963 An equilibrium theory of insular zoogeography. *Evolution* **17**, 373–87.

MacArthur, R.H. and Wilson, E.O. 1967 *The theory of island biogeography*. Monographs in Population Biology Volume 1. Princeton. Princeton University Press.

Macdonald, M.A. and Bolton, M. 2008 Predation of wader nests in Europe. *Ibis* **150** (Suppl. 1), 54–73.

Macdonald, R. and Whelan, J. 1986 Seasonal variation in feeding range and flock structure of the rook (*Corvus frugilegus*) in eastern Ireland. *Ibis* **128**, 540–47.

Mackey, M., Ó Cadhla, O., Kelly, T.C., Aguilar de Soto, N. and Connolly, N. 2004 *Cetaceans and seabirds of Ireland's Atlantic margin. Volume 1. Seabird distribution, density and abundance*. Cork. Coastal and Marine Resource Centre.

Mallet, J. 2005 Speciation in the 21st century. A review of "Speciation" by J.A. Coyne and H. Allen Orr. *Heredity* **95**, 105–9.

Mallet, J. 2008 A century of evolution: Ernst Mayr (1904–2005). Mayr's view of Darwin: was Darwin wrong about speciation? *Biological Journal of the Linnaean Society* **95**, 3–16.

Mallory, J.P. 2013 *The origins of the Irish*. London. Thames and Hudson.

Manne, L.L., Pimm, S.L., Diamond, J.M. and Reed, T.M. 1998 The form of curves: a direct evaluation of MacArthur and Wilson's classic theory. *Journal of Animal Ecology* **67**, 784–94.

Marchant, J.H. and Gregory, R.D. 1999 Numbers of nesting rooks *Corvus frugilegus* in the United Kingdom in 1996. *Bird Study* **46**, 258–73.

Marquet, P.A., Navarette, S.A. and Castilla, J.C. 1995 Body size, population density, and the Energetic Equivalence Rule. *Journal of Animal Ecology* **64**, 325–32.

Marquiss, M. and Booth, C.J. 1986 The diet of Ravens in Orkney. *Bird Study* **33**, 190–5.

Masden, E.A., Haydon, D.T., Fox, A.D. and Furness, R.W. 2010 Barriers to movement: modelling energetic costs of Avoiding Marine Wind Farms Amongst Breeding Seabirds. *Marine Pollution Bulletin* **60**, 1085–91.

Mason, C.F. and Macdonald, S.M. 2004 Distribution of foraging rooks *Corvus frugilegus* and rookeries in a landscape in eastern England dominated by winter cereals. *Folia Zoologica* **53**, 179–88.

Matute, D.R. 2013 The role of founder effects on the evolution of reproductive isolation. *Journal of Evolutionary Biology* **26**, 2299–311.

Mayer-Lucht, Y., Mulder, K.P., James, M.C., McMahon, B.J., Buckley, K., Piertney, S.B. and Hoglund, J. 2016 Adaptive and neutral genetic differentiation among Scottish and endangered Irish red grouse (Lagopus lagopus scotica). *Conservation Genetics* **17**, 615–30.

Mayr, E. 1942 *Systematics and the Origin of Species*. New York. Columbia University Press.

Mayr, E. 1963 *Animal Species and Evolution*. Massachusetts. Harvard University Press.

Mayr, E. 1991 *One long argument – Charles Darwin and the genesis of modern evolutionary thought*. London. Penguin Press.

Mayr, E. 1999 Systematics and the origin of species from the viewpoint of a zoologist. Harvard. Harvard University Press.

Mayr, G. 2011 Metaves, mirandornithes, strisores and other novelties – a critical review of higher-level phylogeny of neornithine birds. *Journal of Zoological Systematics and Evolutionary Research* **49**, 58–76.

McCann, K.S. 2000 The diversity-stability debate. *Nature* **405**, 228–33.

McCarthy, T.K. 1986 Biogeographical aspects of Ireland's invertebrate fauna. In D.P. Sleeman, R. Devoy and P.C. Woodman (eds) *Post Glacial Colonization of Ireland. Occasional Publication of the Irish Biogeographical Society* **1**: 67–81.

McCollin, D. 2014 Abundance estimates for landbirds and seabirds extracted and compiled from annual reports of the Skokholm bird observatory. http://doi.pangaea.de/10.1594/ PANGAEA 833759.

McCollin, D. 2015 The curious case of Skokholm: equilibrium, non-equilibrium and phase shift in an island bird assemblage. *Ecography* **38**, 986–91.

McCollin, D. 2017 Turnover dynamics of breeding landbirds on islands: is island biogeographic theory true but trivial over decadal time scales. *Diversity* **9**, 1–7.

McCune, B. and Grace, J.B. 2002 *Analysis of ecological communities*. Oregon. MjM Software Design.

McGeehan, A. 2014 On the rocks – Ireland's wild rock doves. *Wings* (autumn) 2014, 6–7.

McGeehan, A. 2018 *To the ends of the earth: Ireland's place in bird migration*. Cork. Collins Press.

McGreal, E. 2007 The breeding status of the raven Corvus corax in southwest Mayo. *Irish Birds* **8**, 247–53.

McGreal, E. 2013 Probable breeding by Leach's storm petrel Oceanodroma leucorrhoa at Bills Rocks, County Mayo; results of a census of its breeding birds. *Irish Birds* **9**, 636–8.

McGuinness, K.A. 1984 Equations and explanations in the study of species-area curves. *Biological Reviews* **59**, 423–40.

McKay, B.D. and Zink, R.M. 2010 The causes of mitochondrial DNA tree paraphyly in birds. *Molecular Phylogenetics and Evolution* **54**, 647–50.

McKechnie, A.F. 2015 Global variation in avian metabolic rates and the slow pace of life of tropical birds. *Functional Ecology* **29**, 297–8.

McLoughlin, D.T. 2011 Management prescriptions for twite in Ireland. Irish Wildlife Manuals No. 52. Dublin. National Parks and Wildlife Service.

McLoughlin, D.T. and Cotton, D. 2008 The status of the twite Carduelis flavirostris in Ireland 2008. *Irish Birds* **8**, 323–30.

McMahon, B.J., Johansson, M.P., Piertney, S.B., Buckley, K. and Hoglund, J. 2012 Genetic variation among endangered Irish red grouse (*Lagopus lagopus hibernicus*) populations: implications for conservation and management. *Conservation Genetics* **13**, 639–47.

Mettke-Hofmann, C. 2017 Avian movements in a modern world: cognitive challenges. *Animal Cognition* **20**, 77–86.

Mitchell, P.I., Newton, S.F., Ratcliffe, N. and Dunn, T.E. 2004 *Seabird populations of Britain and Ireland*. London. Poyser.

Moorcroft, D., Wilson, J.D. and Bradbury, R.B. 2006 Diet of nestling Linnets *Carduelis cannabina* on lowland farmland before and after agricultural intensification. *Bird Study* **53**, 156–62.

Moya, A., Galiana, A. and Ayala, F.J. 1995 Founder effect speciation theory: failure of experimental corroboration. *Proceedings of the National Academy of Sciences* **92**, 3983–6.

Mullarney, K., Svensson, L., Zetterstrom, D. and Grant, P.J. 2001 Collins bird guide. London. Harper Collins.

Munns, D.J. 1956 Further notes on the birds of the Blasket Islands. *Bird Study* **3**, 248–50.

Munro, M. 2017 What's killing the World's shorebirds? *Nature* **541**, 16–20.

Murray, T., Clotworthy, C. and Bleasdale, A. 2013 A survey of red grouse (*Lagopus lagopus scoticus*) in the Owenduff/Nephin complex Special Protection Area, County Mayo. Irish Wildlife Manual No. 77. Dublin. National Parks and Wildlife Service.

Nairn, R. and O'Halloran, J. (eds) 2012 *Bird Habitats in Ireland*. Cork. Collins Press.

Neubauer, P., Jensen, O.F., Hutchings, J.A. and Baum, J.K. 2013 Resilience and recovery of overexploited marine populations. *Science* **340**, 347–9.

Newton, I. 1967 The adaptive radiation and feeding ecology of some British finches. *Ibis* **109**, 33–98.

Newton, I. 2003 *The speciation and biogeography of birds*. London. Academic Press.

Newton, S.F. 2012 Sea cliffs, islands and rocky coasts. In R. Nairn and J. O'Halloran (eds) *Bird habitats in Ireland*, 138–51. Cork. Collins Press.

Newton, S.F., Harris, M.P. and Murray, S. 2015a Census of gannet Morus bassanus colonies in Ireland in 2013–2014. *Irish Birds* **10**, 215–20.

Newton, S.F., Lewis, L. and Trewby, M. 2015 Results of a breeding survey of important cliff-nesting seabird colonies in Ireland 2015 – with an interim analysis of population changes. Unpublished Report. Kilkcoole. Birdwatch Ireland.

Newton, S.F., Moralee, A. and Cabot, D.B. 2002 Great skua Catharacta skua breeding in Ireland. *Irish Birds* **7**, 129.

Nichols, J.D. 1991 Responses of North American duck populations to exploitation. In Perrins, C.M., Lebreton, J-D. and Hirons, G.J.M. (eds) *Bird population studies*, 498–525. Oxford Ornithology Series 1. Oxford. Oxford University Press.

Nuorteva, P. 1971 The synanthropy of birds as an expression of the ecological cycle disorder caused by urbanisation. *Annales Zoologici Fennici* **8**, 547–53.

Ó Fiannachta, P. 1983 An tOileánach: The Islandman by Tomas Ó Criomhthain. In P. Ó Conluain (ed.) *Islands and authors. Pen pictures of life past and present on the islands of Ireland*. 44–58. Cork. Mercier Press.

O'Connell, S., Irwin, S., Wilson, M.W., Sweeney, O., Kelly, T.C. and O'Halloran, J. 2012 How can forest management benefit bird communities? Evidence from eight years of research in Ireland. *Irish Forestry* **69**, 44–57.

O'Halloran, J., Nairn, R., Smiddy, P. and Crowe, O. 2012 Bird habitats: a synthesis and future perspectives. In R. Nairn and J. O'Halloran (eds) *Bird habitats in Ireland*. 247–68. Cork. Collins Press.

O'Regan, S.M., Flynn, D., Kelly, T.C., O'Callagahan, M.J.A., Pokrovskii, A. and Rachinskii, D. 2012 The response of the Woodpigeon (*Columba palumbus*) to relaxation of intraspecific competition: a hybrid modelling approach *Ecological Modelling* **224**, 54–64.

Oksanen, J., Blanchet, F.F, Kindt, R., Legendre, P. and Minchin, P.R. 2016 Vegan Community Ecology. R package version.

Olea, P.P. 2009 Analysing spatial and temporal variation in colony size: an approach using autoregessive mixed models and information theory. *Population Ecology* **51**, 161–74.

Olea, P.P. and Baglione, V. 2008 Population trends of rooks *Corvus frugilegus* in Spain and the importance of refuse tips. *Ibis* **150**, 98–109.

Olson, V.A., Davies, R.G., Orme, C.P.L., Thomas, G.H., Meiri, S., Blackburn, T.M., Gaston, K.J., Owens, I.P.F. and Bennett, P.M. 2009 Global biogeography and ecology of body size in birds. *Ecology Letters* **12**, 249–59.

Orłowski, G. and CZapulak, A. 2007 Different extinction risks of the breeding colonies of rooks *Corvus frugilegus* in rural and urban areas of SW Poland. *Acta Ornithologica* **42**, 145–55.

Owens, I.P.F., Scott, S.N., Robinson, S.I., Clegg, S.M. and Kikkawa, J. 2006 Large body size in island-dwelling passerines: the roles of insular specialisation, niche expansion and ecological release. *Acta Zoologica Sinica* **52** (Supplement) 262–6.

Paradis, E., Baillie, S.R., Sutherland, W.J. and Gregory, R.D. 1998 Patterns of natal and breeding dispersal in birds. *Journal of Animal Ecology* **67**, 518–36.

Parkin, D.T. and Knox, A.G. 2010 *The status of birds in Britain and Ireland*. London. Christopher Helm.

Passy, S.J. 2012 A hierarchical theory of macroecology. *Ecology Letters* **15**, 923–4.

Peach, W.J., Thompson, P.S. and Coulson, J.C. 1994 Annual and long term variation in the survival rates of lapwings *Vanellus vanellus Journal of Animal Ecology* **63**, 60–70.

Pennekamp, F., Pontarp, M., Tabi, A., Altermatt, F., Alther, R., Choffat, Y., Fronhofer, E.A., Ganesanandamoorthy, P., Garnier, A., Griffiths, J.I., Greene, S., Horgan, K., Massie, T.M., Mächler, E., Palamara, G.M., Seymour, M. and Petchey, O.L. 2018 Biodiversity increases and decreases ecosystem stability. *Nature* **563**, 109–12.

Peron, G. 2013 Compensation and additivity of anthropogenic mortality: life history effects and review of methods. *Journal of Animal Ecology* **82**, 408–17.

Perrins, C.M. 1985 The effect of man on the British avifauna. In J.L. Bunning (ed.) *Proceedings of the Birds and Man symposium*. 5–29. Johannesburg.

Pethica, J.L. and Roy, J.C. (eds) 1998 *'To the land of the free from this island of slaves'. Henry Stratford Persse's letters from Galway to America, 1821–1832*. Cork. Cork University Press.

Phillips, R.A., Gales, R., Baker, G.B., Double, M.C., Favero, M., Quintina, F., Tasker, M.L., Weimerskirch, H., Uhart, M. and Wolfaarrdt, A. 2016 The conservation status and priorities for albatrosses and large petrels. *Biological Conservation* **201**, 169–83.

Pierce, S. and Roe, J. 2006 Merlins Falco columbarius hunting at a small marine island. *Irish Birds* **8**, 79–84.

Pilcher, J. and Hall, V. 2001 *Flora Hibernica*. Cork. Collins Press.

Pimm, S.L. and Raven, P. 2000 Extinction by numbers. *Nature* **403**, 843–45.

Pimm, S.L., Diamond, J., Reed, T.M., Russell, G.J. and Verner, J. 1993 Times to extinction for small populations of large birds. *Proceedings of the National Academy of Sciences, USA* **90**, 10871–5.

Pollock, C.M. and Barton, C. 2008 A gap analysis of Irish waters using the European Seabirds at Sea database. Irish Wildlife Manuals Report. Dublin. National Parks and Wildlife Service.

Pollock, C.M., Reid, J.B., Webb, A. and Tasker, M.L. 1997 The distribution of seabirds and cetaceans in the waters around Ireland. Report No. 267. Aberdeen. Joint Nature Conservation Committee.

Praeger, R.L. 1911 Clare Island Survey Part 10 Phanerogamia and Pteridophyta. *Proceedings of the Royal Irish Academy* **31**, 1–112.

Praeger, R.L. 1915 General introduction and narrative. *Proceedings of the Royal Irish Academy* **31**(1911–1915), Section 1, Part 1 (1915), 1–3.

Preston, K. 1979 Twenty sixth Irish bird report 1978. *Irish Birds* **1**, 413–99.

Prop, J. and Black, J.M. 1998 Food intake, body reserves and reproductive success of Barnacle Geese *Branta leucopsis* staging in different habitats. *Norsk Polarinstitutt Skrifter* **200**, 175–93.

Prum, R.O., Berv, J.S., Dornberg, A., Field, D.S., Townsend, J.F., Lemmon, T.M. and Lemmon, A.R. 2015 A comprehensive phylogeny of birds (Aves) using targeted next generation DNA sequencing. *Nature* **526**, 569–72.

Quammen, D. 1996 *The song of the Dodo*. London. Hutchinson.

Ratcliffe, D. 1990 *Bird life of mountain and woodland*. Cambridge. Cambridge University Press.

Ratcliffe, D. 1997 *The Raven*. London. Poyser.

Reed, T. 1980 Turnover frequency in island birds. *Journal of Biogeography* **7**, 329–35.

Reed, T. 1981 The number of breeding landbird species on British islands. *Journal of Animal Ecology* **50**, 613–24.

Reuman, D.C., Mulder, C., Benasek-Richter, C., Blandanier, M-F.C., Breure, A.M., Hollander, H.D., Knietal, J.M., Rafaelli, D., Woodward, G. and Cohen, J.E. 2009 Allometry of body size and abundance in 166 food webs. *Advances in Ecological Research* **41**, 1–44.

Ricklefs, R.E. and Bermingham, E. 2001 Non equilibrium dynamics of the Lesser Antillian avifauna. *Science* **294**, 1522–24.

Riddle, B.R. and Hafner, D.J. 1999 Species as units of analysis in ecology and biogeography: time to take the blinkers off. *Global Ecology and Biogeography* **8**, 433–41.

Ridley, M. 2004 *Evolution*. Third edition. Oxford. Blackwell.

Rivelan, P., Barbaud, C., Inchausti, P. and Weimerskirch, H. 2010 Combined impacts of longline fisheries and climate change on the persistence of the Amsterdam albatross Diomedea amsterdamensis. *Ibis* **152**, 6–18.

Rolland, V., Weimerskirch, H. and Barbaud, C. 2010 Relative influence of fisheries and climate change on the demography of four albatrosses. *Global Change Biology* **16**, 1910–22.

Rosa, I.M.D., Marques, A.M., Palmunka, G.M., Costa, H., Mascarenhas, M., Fonesca, C. and Bernardino, J. 2016 Classification success of six machine learning algorithms in radar ornithology. *Ibis* **158**, 28–42.

Rosenzweig, M.L. and Ziv, Y. 1999 The echo pattern of species diversity: pattern and process. *Ecography* **22**, 614–28.

Russell, D.J.F., Wanless, S., Collingham, Y.C. and Hamer, K.C. 2015 Predicting future European breeding distributions of British seabird species under climate change and unlimited no dispersal scenarios. *Diversity* **7**, 342–59.

Russell, G.J., Diamond, J.M., Pimm, S.L. and Reed, T.M. 1995 A century of turnover: community dynamics on three timescales. *Journal of Animal Ecology* **64**, 628–41.

Russell, G.J., Diamond, J.M., Reed, T.M. and Pimm, S.L. 2006 Breeding birds on small islands: island biography or optimal foraging? *Journal of Animal Ecology* **75**, 324–39.

Ruttledge, R.F. 1947 Report on birds at Irish Light-stations 1946. *Irish Naturalists' Journal* **9**, 57–61.

Ruttledge, R.F. 1950 A list of the birds of the Counties of Galway and Mayo: showing their status and distribution. *Proceedings of the Royal Irish Academy* **52B**, 315–81.

Ruttledge, R.F. 1957 The birds of Inishbofin, Co. Galway. *Bird Study* **4**, 71–80.

Ruttledge, R.F. 1963 Migrant and other birds of Great Saltee, Co. Wexford. *Proceedings of the Royal Irish Academy* **63B**, 71–86.

Ruttledge, R.F. 1966 *Ireland's birds*. London. Witherby.

Ruttledge, R.F. 1975 *A list of the birds of Ireland*. Dublin. National Museum of Ireland.

Ruttledge, R.F. 1994 *Birds in counties Galway and Mayo*. Dublin. Irish Wildbird Conservancy.

Ruttledge, R.F. and Oglivie, M.A. 1979 The past and current status of the Greenland white-fronted goose in Ireland and Britain. *Irish Birds* **1**, 293–363.

Ryle, T. 2013 Vegetation-Environment interactions on Clare Island. In D. Synnott (ed.) *New Survey of Clare Island. Volume 7: Plants and fungi*, 27–178. Dublin. Royal Irish Academy.

Saastamoinen, M., Bocedi, G., Cote, J., Legrand, O., Guillome, F., Wheat, C.W., Fronhofer, E.A., Garcia, C., Henly, R., Husby, A., Baguette, M., Bonte, D., Coulon, A., Kokko, H., Mathyssen, E., Niitepõld, K., Nonaka, E., Stevens, V.M., Travis, J.M.J., Donoghue, K., Bullock, J.M. and del Mar Delgado, M. 2018 Genetics of dispersal. *Biological Reviews* **93**, 574–99.

Saether, B.E. and Bakke, O. 2000 Avian life history variation and contribution of demographic traits. *Ecology* **81**, 642–53.

Sage, B. and Whittington, P.A. 1985 The 1980 sample survey of rookeries. *Bird Study* **32**, 77–81.

Sandercock, B.K. 2003 Estimation of survival rates for wader populations: a review of mark-recapture methods. *Wader Study Group Bulletin* **100**, 163–74.

Sangster, G. 2014 The application of species criteria in avian taxonomy and its implications for the debate of species concepts. *Biological Reviews* **89**, 199–214.

Sangster, G. 2018 Integrative taxonomy of birds: the nature and delimitation of species. In T. Tietze (ed.) *Bird species, how they arise, modify and vanish*. 9–38. Switzerland. Springer.

Saunders, H. 1899 *Manual of British birds*. London. Gurney and Jackson.

Savory, C.J. 1978 Food consumption of red grouse in relation to the age and productivity of heather. *Journal of Animal Ecology* **47**, 269–82.

Schneider, N.A., Low, M., Arlt., D. and Part, T. 2012 Contrast in edge vegetation structure modifies the predation risk of natural ground nests in an agricultural landscape. *PloS One* **7**, e31517.doi 10.1371/journal.pone.00131517.

Scott, S.N., Clegg, S.M., Blomberg, S.P., Kikkawa, J. and Owens, I.P.F. 2003 Morphological shifts in island-dwelling birds: the roles of generalist foraging and niche expansion. *Evolution* **57**, 2147–56.

Sellers, R.M. 2006 Breeding population estimate for Northern Wheatears in Britain. *British Birds* **99**, 533–35.

Semenov, G.A., Koblik, E.A., Red'Kin, Y.A. and Badyaev, A.V. 2018 Extensive phenotypic diversification coexists with little genetic divergence and lack of population structure in the white wagtail subspecies complex. *Journal of Evolutionary Biology* doi:10.1111/jeb.13305.

Seward, A.M., Beale, C.M., Gilbert, L., Hefin Jones, T. and Thomas, R.J. 2013 The impact of increased food availability on survival of a long-distance migratory bird. *Ecology* **94**, 221–30.

Sharpe, C.M. 2007 *Manx bird atlas*. Liverpool. Liverpool University Press.

Sharrock, J.T.R. 1973 *The natural history of Cape Clear Island*. Berkhamsted. Poyser.

Sharrock, J.T.R. 1976 *The atlas of breeding birds in Britain and Ireland*. Tring. British Trust for Ornithology.

Sharrock, J.T.R. and Fogden, S.C.L. 1965 The breeding passerines of Cape Clear 1965. Cape Clear Bird Observatory Report **7**, 42–49.

Sibley, R.M., Barker, D., Hone, J. and Pagel, M. 2007 On the stability of populations of mammals, birds, fish and insects. *Ecology Letters* **10**, 970–6.

Simberloff, D. 1974 Equilibrium theory of island biogeography and ecology. *Annual Review of Ecology and Systematics* **5**, 161–82.

Simberloff, D. 1976 Experimental zoogeography of islands: effects of island size. *Ecology* **57**, 629–48.

Sleeman, D.P., Carlsson, J. and Carlsson, J.E.L. (eds) 2014 *Mind the gap II: new insights into the Irish postglacial*. Belfast. Irish Naturalists' Journal.

Sleeman, D.P., Devoy, R. and Woodman, P.C. (eds) 1986 *Post Glacial Colonization of Ireland*. Occasional Publication of the Irish Biogeographical Society **1**.

Smiddy, P. 2010 A systematic list of Irish Birds. *Irish Birds* **9**, 130–4.

Smiddy, P. 2014 Breeding birds in Ireland: success and failure among colonists. In D.P. Sleeman, J. Carlsson and J.E.L. Carlsson (eds) *Mind the gap II: new insights into the Irish postglacial*. 89–99. Belfast. Irish Naturalists' Journal.

Smiddy, P. 2016 Diet of coastal breeding Ravens *Corvus corax* in east County Cork. *Irish Birds* **10**, 335–8.

Smiddy, P., O'Halloran, J., O'Mahony, B. and Taylor, A.J. 1995 The breeding biology of the dipper *Cinclus cinclus* in southwest Ireland. *Bird Study* **42**, 76–81.

Southwood, T.R.E., Brown, V.K., Reader, P.M. and Green, E.E. 1986 The use of different stages of succession by birds. *Bird Study* **33**, 159–63.

Stephens, P. and Sutherland, W. 1999 Consequences of the Allee effect for behaviour, ecology and conservation. *Trends in Ecology and Evolution* **14**, 401–5.

Storch, D. and Gaston, K.J. 2004 Untangling ecological complexity on different scales of space and time. *Basic and Applied Ecology* **5**, 389–400.

Stracey, C.M. and Pimm, S.L. 2009 Testing island biogeography theory with visitation rates of birds to British islands. *Journal of Biogeography* **36**, 1532–9.

Suddaby, D., Nelson, T. and Veldman, J. 2010 Resurvey of breeding wader populations of machair and associated wet grasslands in north-west Ireland. *Irish Wildlife Manuals* No. 44. Dublin. National Parks and Wildlife Service.

Sutherland, W.J. *et al.* 2013 Identification of 100 fundamental ecological questions. *Journal of Ecology* **161**, 58–67.

Sweeney, O.F., Kelly, T.C., Irwin, S., Wilson, M.W. and O'Halloran, J. 2012 Woodlands, forest and scrub. In Nairn, R. and O'Halloran, J. (eds) *Bird habitats in Ireland*. 105–23. Cork. Collins Press.

Sweeney, O.F., Wilson, M.W., Irwin, S., Kelly, T.C. and O'Halloran, J. 2010 Breeding bird communities of second-rotation plantations at different stages of the forest cycle. *Bird Study* **57**, 301–14.

Taylor, A.J. and O'Halloran, J. 2002 The decline of the corn bunting Milaria calandra in the Republic of Ireland. *Biology and Environment: Proceedings of the Royal Irish Academy* **102B**, 165–75.

Taylor, C.M. and Hastings, A. 2005 Allee effects and biological invasions. *Ecology Letters* **8**, 895–908.

Thom, V.M. 1986 *Birds in Scotland*. Calton. Poyser.

Thomas, L., Buckland, S.T., Burnham, K.P., Anderson, D.R., Laake, J.L., Borchers, D.L. and Strindberg, S. 2013 Distance sampling. *Encyclopedia of Environmetrics*. Second edition. Doi: 10.1002/9780470057339.vad033.pub2

Thomas, L., Buckland, S.T., Rexstad, E.A., Laake, J.L., Strindberg, S., Hedley, S.L., Bishop, J.R.B., Marques, T.A. and Burnham, K.P. 2010 Distance software: design and analysis of distance sampling surveys for estimating population size. *Journal of Applied Ecology* **47**, 5–14.

Thomson, D.L., Baillie, S.R. and Peach, W.J. 1997 The demography and age specific annual survival of Song Thrushes during periods of population stability and decline. *Journal of Animal Ecology* **66**, 414–24.

Thorup, O. 2006 *Breeding waders in Europe 2000*. International Wader Studies 14. United Kingdom. International Wader Study Group.

Tietze, T. (ed.) 2018 *Bird species, how they arise, modify and vanish*. Switzerland. Springer.

Tilman, D. 1996 Biodiversity: population versus ecosystem stability. *Ecology* **77**, 350–63.

Toews, D.P.L. 2015 Biological species and taxonomic species: will a new null hypothesis help? (A comment on Gill, 2014). *The Auk* **132**, 78–81.

Triantis, K.A., Guilhaumon, F. and Whitaker, R.J. 2012 The island species area relationship: biology and statistics. *Journal of Biogeography* **39**, 215–31.

Udvardy, M.D.F. and Papp, C.S. 1969 *Dynamic zoogeography*. London. Von Nostrand Reinhold Company.

Ussher, R.J. 1912 A biological survey of Clare Island in the county of Mayo, Ireland, and adjoining district. *Proceedings of the Royal Irish Academy* **31**, Section 11, Part 20.

Ussher, R.J. and Warren, R. 1900 *Birds of Ireland*. London. Gurney and Jackson.

Uva, V., Packert, M., Cibois, A., Fumagalli, L. and Roulin, A. 2018 Comparative molecular phylogeny of the barn owls and relatives (Family: Tytonidae) and their six major pleistocene radiations. *Molecular Phylogenetics and Evolution* **125**, 127–32.

Uy, J.A.C., Irwin, D.E. and Webster, M.S. 2018 Behavioral isolation and incipient speciation in birds. *Annual Review of Ecology, Evolution and Systematics* **49**, 1–24.

Valone, T.J. 2007 From eavesdropping on performance to copying the behaviour of others: a review of public information use. *Behavioral Ecology and Sociobiology* **62**, 1–14.

Venables, L.S.V. 1973 Rooks in Shetland. *Scottish Birds* **7**, 406.

Venables, L.S.V. and Venables, U.V. 1953 Rooks nesting in Shetland. *British Birds* **46**, 265–66.

Venkatraman, M.X., Deraad, D.A., Tsai, W.L.E., Zarza, E., Zellmer, A.J., Maley, J.M. and McCormack, J.E. 2019 Cloudy with a chance of speciation: integrative taxonomy reveals extraordinary divergence within a Mesoamerican cloud forest bird. *Biological Journal of the Linnaean Society* **126**, 1–15.

Vorisek, P., Jiguet, F., van Strien, A., Skorpilova, J., Kvalanova, A. and Gregory, R.D. 2010 Trends in abundance and biomass of widespread European farmland birds: how much have we lost? *European trends in farmland birds. BOU Proceedings Lowland Farmland Birds III*.

Votier, S.C., Furness, R.W., Bearhop, S., Crane, J.E., Caldow, R.W.G., Catry, P., Ensor, K., Hamer, K.C., Hudson, A.V., Kalmbach, E., Klomp, N.I., Pfeiffer, S., Phillips, R.A., Prieto, I. and Thompson, D.R. 2004 Changes in fisheries discard rates and seabird communities. *Nature* **427**, 727–30.

Vullings, W. Collins, J.F. and Smillie, G. 2013 *New Survey of Clare Island. Volume 8: Soils and soil associations*. Dublin. Royal Irish Academy.

Wallace, A.R. 1880 *Island Life: or the causes of insular faunas and floras, including a revision and attempted solution of the problem of geological climates*. London. Macmillan.

Wallace, D.I.M., McGeehan, A. and Allen, D. 2001 Autumn migration in westernmost Donegal. *British Birds* **94**, 103–20.

Walsh, A. and Merne, O.J. 1988 Barnacle geese in Ireland, spring 1988. *Irish Birds* **5**, 539–50.

Walsh, P.M., Halley, D.J., Harris, M.P., del Nevo, A., Sim, I.M.W. and Tasker, M.L. 1995 Seabird monitoring handbook for Britain and Ireland. Peterborough. JNCC/ RSPB/ITE/ Seabird Group.

Warren, B.H., Simberloff, D., Ricklefs, R.E., Aguile, R., Condamine, F.L., Gravel, D., Morlon, H., Mouquet, N., Rosindell, J., Casquet, J., Cornault, J., Fernandez-Palacios, J.M., Hengl, T., Norder, S.J., Rijsdijk, K.F., Sanmartin, D., Triantis, K.A., Valente, L.M., Whittaker, R.J., Gillespie, R.G., Emerson, B.C. and Thebaud, C. 2015 Islands as model systems in ecology and evolution: prospects fifty years after MacArthur-Wilson. *Ecology Letters* **18**, 200–17.

Weigelt, P., Steinbauer, M.J., Cabral, J.S. and Kreft, H. 2016 Late Quaternary climate change shapes island biodiversity. *Nature* **532**, 99–102.

Wernham, C., Toms, M., Marchant, J., Clark, J., Siriwardina, G. and Baillie, S. (eds) 2002 *The migration atlas: movement of the birds of Britain and Ireland*. London. Poyser.

Whilde, A. 1979 Auks trapped in salmon drift nets. *Irish Birds* **1**: 370–6.

Whittaker, R.J. and Fernandez-Palacios, J.M. 2007 *Island biography*. Second edition. Oxford. Oxford University Press.

Whittaker, R.J., Fernandez-Palacios, J.M., Matthews, T.J., Borregaard, M.K. and Triantis, K.A. 2017 Island biogeography: taking the long view of nature's laboratories. *Science* **357**. DOI: 10.1126/science.aam8326.

Whittaker, R.J., Triantis, K.A. and Ladle, R.J. 2008 A general dynamic theory of oceanic island biogeography. *Journal of Biogeography* **36**, 977–94.

Whittaker, R.J., Triantis, K.A. and Ladle, R.J. 2010 A general dynamic theory of oceanic island biogeography: extending the MacArthur and Wilson theory to accommodate the rise and fall of volcanic islands. In J.B. Losos and R. Ricklefs (eds) *The theory of island biogeography revisited*. 88–115. Princeton. Princeton University Press.

Whittow, G.C. 2002 Seabird reproductive physiology and energetics. In E.A. Schreiber and J. Burger (eds) *Biology of marine birds*, 409–37. London, Boca Raton, New York and Washington. CRC Press.

Williams, J.B., Miller, R.A., Harper, J.M. and Wiersma, P. 2010 Functional linkages for the pace of life, life-history and environment in birds. *Integrative and Comparative Biology* **50**, 855–68.

Williamson, K. 1975 Birds and climate change. *Bird Study* **22**, 143–64.

Williamson, M. 1981 *Island populations*. Oxford. Oxford University Press.

Williamson, M. 1983 The land-bird community of Skokholm: ordination and turnover. *Oikos* **41**, 378–84.

Williamson, M. 1989 The MacArthur – Wilson theory today: true but trivial. *Journal of Biogeography* **16**, 3–4.

Wilson, M.W., Gittings, T., Pithon, J., Kelly, T.C., Irwin, S. and O'Halloran, J. 2012 Bird diversity of afforestation habitats in Ireland: current trends and likely impacts. *Biology and Environment: Proceedings of the Royal Irish Academy* **112B**, 55–68.

Wilson, M.W., Pithon, J., Gittings, T., Kelly, T.C., Giller, P.S. and O'Halloran, J. 2006 Effects of growth stage and tree species composition on breeding bird assemblages of plantation forests. *Bird Study* **53**, 225–36.

Winstanley, D.R., Spencer, R.G. and Williamson, K. 1974 Where have all the whitethroats gone? *Bird Study* **24**, 1–14.

Winters, P. 2006 *A summer bird survey of Clare Island*. Unpublished MSc thesis. Dublin. Centre for the Environment, Trinity College.

Woodward, F.I., Quaife, T. and Lomas, M.R. 2010 Changing climate and the Irish landscape. *Biology and Environment: Proceedings of the Royal Irish Academy* **110B**, 1–16.

Wynne-Edwards, V.C. 1964 Oceanic birds. In A.L. Thomson (ed.) *A new dictionary of birds*. London. Nelson.

Yom-Tov, Y. 1974 The effect of food and predation on breeding density and success, clutch size, and laying date of the crow (*Corvus corone* L.). *Journal of Animal Ecology*, **43**, 479–98.

Young, G., Lawlor, M., Paintin, T., Atkinson, M. and Watson, J. 2014 *A working list of the birds of the Channel Islands*. www.jerseybirds.co.uk. Accessed 19 September 2016.

TAXONOMIC INDEX

Page references in bold indicate tables. Those in italics indicate illustrations.

INDEX

Page references in bold indicate tables.
Those in italics indicate illustrations.